T0188134

The Sudarium of Oviedo

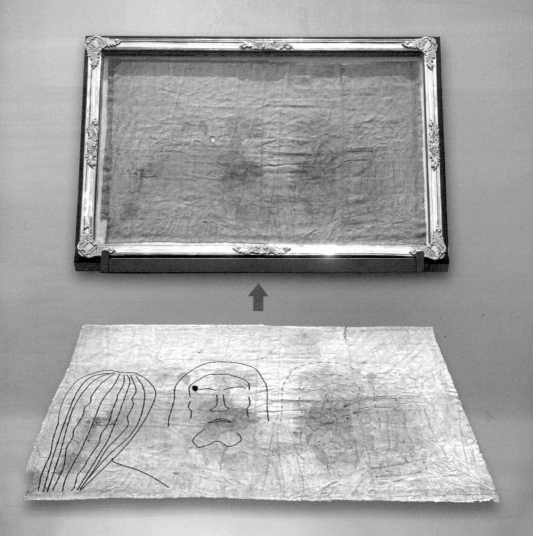

Jenny Stanford Series on Christian Relics and Phenomena

Series Editor
Giulio Fanti

Titles in the Series

Jenny Stanford Series on Christian Relics and Phenomena — Volume 4

The Sudarium of Oviedo
Signs of Jesus Christ's Death

César Barta
Series Editor: Giulio Fanti

JENNY STANFORD
PUBLISHING

Published by

Jenny Stanford Publishing Pte. Ltd.
Level 34, Centennial Tower
3 Temasek Avenue
Singapore 039190

Email: editorial@jennystanford.com
Web: www.jennystanford.com

British Library Cataloguing-in-Publication Data
A catalogue record for this book is available from the British Library.

The Sudarium of Oviedo: Signs of Jesus Christ's Death

ISBN 978-981-4968-13-3 (Hardcover)
ISBN 978-1-003-27752-1 (eBook)

To all those
who look for the truth without prejudices... if any.

Contents

Foreword

The Centro Español Sindonología (CES)
and the Holy Sudarium of the Cathedral of Oviedo

By Jorge Manuel Rodríguez Almenar, President of the CES

I know César Barta since the 90s of the past century, and we have worked in close association. I appreciate his proposal to write the foreword to his work, in my position as the president of the *Centro Español de Sindonología* (Spanish Center of Sindonology, CES), and I agree gladly. I think my contribution to his book should focus on the story of the beginning of the research carried out in the recent decades on the "Holy Sudarium", because it is necessary that the reader does not think that what is stated in this book is the result of mere individual lucubration. What here is described is largely the result of a collective investigation of many years and—as this book shows—does not come from nothing but has a history and is derived from particular circumstances that I need to remark.

I. The Holy Sudarium as the Object of the CES Study

By "*sudarium*", we mean a relic supposedly preserved in the capital of the Principality of Asturias (autonomous community in northern Spain), which has been called for centuries "*Sudarium Domini*" or "*Holy Sudarium*" while nobody could explain so far why exactly it received that denomination. It has been also known as "*Holy Face*", although it has no image. On the contrary, it seems rather a dirty cloth, stained and wrinkled, with apparent worthlessness. However, it has been traditionally linked to the small cloth mentioned in the Gospel of John in the sepulcher of Jesus (Jn 20: 7): "*the sudarium that had been on his head*". In 1987, a cultural association was created in Valencia, but for a national scope, called Centro Español Sindonología (CES). Its founder, the writer Ms Manuela Corsini Ordeig was a *most beloved* member of the International Center for Syndonology of Turin. And our association, as its name explains[1], had as its initial objective the dissemination of studies on the so-called *Holy Shroud* of Turin, and also the possible direct or indirect research about every aspect related to the Turin relic.

I was fortunate to be part of the initial group of young university students who supported—and almost pushed—Manuela Corsini to create the association. At that time, virtually none of us knew of the existence of the so-called Holy Sudarium. In one of the initial meetings of the members, our founder spoke to us about a phone call received from an engineer of Madrid (Guillermo Heras), who spoke about the need of including among our objectives the study of the relic that was in the cathedral of Oviedo. The subject seemed to be an interesting topic, but at that time we did not have the necessary infrastructure to start such direct investigation. The most that we could refer to was what was already known about the Shroud, which was very little.

Guillermo Heras called again our head office few months later. However, in the meantime, the sudden death of our founder happened, and the new president, Celestino Cano Tello, had just taken possession. Fortunately for all, for his work, Guillermo had to move to Valencia frequently in the following months and was reeling off his ideas to the new Board of Directors during those trips. He proposed to form a scientific team to directly study the Holy Sudarium. Guillermo's involvement was providential, because he directly knew the people who had to be addressed in the archbishopric of Oviedo. This fact came for having worked for years near the Asturian capital. He had a clear idea of what kind of people could begin the investigation.

Needless to say, given his proven solvency and ability, Guillermo was appointed Chief of Scientist Section of the CES, and in the following months, EDICES ("Team Research of CES" for its Spanish acronym) was launched.

After the appropriate procedures, on 9 November 1989, the CES obtained the official authorization by the archbishopric of Oviedo and the cathedral chapter to carry out a complete and multidisciplinary study of the *Holy Sudarium*.

---

[1]Sindonología is a word formed by the union of two Greek terms: *sindon* meaning "linen cloth" and *logy that means "study or treaty".*

Our association, which had arisen modestly to extend what was known of the Holy Shroud of Turin, providentially found with a huge field of study that no one had foreseen initially.

This new situation allowed us to broaden our horizons and focus on the study of the *Shroud* its authentic subject context, which actually is the study of the historicity of Jesus of Nazareth. This defined our purpose much better, since we did not intend to support or refute the authenticity of the cloth but rather understand those objects which are *documents* from a secular point of view. Those sources could give us direct or indirect information about the Character that interested us: Jesus himself. One of these objects was the Shroud, but it was not the only one[2]. Therefore, the enlargement of our purposes turned out to be quite suitable because the Sudarium of Oviedo had already been linked for years to the well-known Shroud of Turin, as we shall see.

II. Scientific Investigations on the Sudarium of Oviedo

A. Individual Studies

The discoverer of the importance of the Holy Sudarium was Msgr. Giulio Ricci[3]. Ricci approached it thinking that it could be the *soudarion* (handkerchief) that quotes the Gospel of John (Jn 20,7), but he had understood the evangelical reference in the sense that the aforementioned cloth was a simple chin guard, which would have surrounded the head, to close the mouth of the executed.

When Ricci visited Oviedo, with the intention of learning more about the Sudarium, he was surprised to see a fabric that had nothing to do with what he imagined. Furthermore, it had much more blood on it than the Shroud. It is not surprising that he was puzzled, since the Sudarium had been in Oviedo for a thousand years with no one having understood what it was,

---

[2]Since the modification of the statutes of the CES, in 1994, it is expressly also cited within the scope of study of the association "the Holy Sudarium of Oviedo, the Holy Chalice of Valencia and other relics attributed to Jesus Christ" as long as they can be object of scientific study.

[3]One priest researcher of the Shroud of the Roman Center of Sindonology who worked in the Vatican.

although the weight of tradition had caused it to be saved and venerated.

Perhaps it was never understood because the faithful hoped to see a face and there was no face to see[4]. In addition, the cloth always had been shown vertically (with its larger sides along the vertical). Ricci placed the Sudarium horizontally for the first time and he realized that the two main stains on the cloth had their axis of symmetry in a fold. This was the first factual data verified on the Sudarium and the first step to decipher it. This axis of symmetry indicates that the main stains were formed while the cloth was folded over the source of the bloody liquid.

Ricci could not progress much further because he was wrong in interpreting the sense of the mentioned fold: Instead of bending the surfaces that were in contact at the time of staining, he folded the cloth incorrectly and therefore, only the main stains fitted each other but not the non-symmetrical ones.

However, Ricci, because of the generosity of the canon chapter of the Cathedral of Oviedo, came to Rome with the backing lining that the Sudarium had had for centuries and obtained permission to cut a small portion of the Sudarium, which he took.

Sometime later, two researchers were endorsed by Ricci to go to Oviedo and take samples. They were Prof. Baima-Bollone from the University of Turin (also director of the International Center for *Sindonología* at that time) and the Swiss palynologist Max Frei, who prepared a report on the pollens identified in the Sudarium of Oviedo and compared them with those of the Shroud of Turin. Frei concluded that the only plant species that had left pollen on both cloths were those that were linked to Palestine.

B. The EDICES Study

However, as I said, since 1989 it is EDICES ("CES Research Team"), on behalf of the CES, which has been conducting a specific multidisciplinary study on the cloth, fulfilling the official

---

[4]Ambrosio Morales, in his memorial about Spain for King Philip the second, reaching the Holy Chamber and describing the Sudarium (which he calls Holy Face), says that it is affirmed that there is a face there, but, because of his many sins", he is not able to see it.

commission of the CES. EDICES is an international team, led by the engineer and lawyer Guillermo Heras Moreno until 2012 and later by the forensic doctor Alfonso Sánchez Hermosilla, both including as deputy coordinator the chemist Montero Felipe Ortego.

Along the years, EDICES has integrated about 30 or 35 members: physicist (as César Barta himself), chemists, forensic doctors, biologists, botanists, forensic anthropologists,..., members of various prestigious public and private entities in Spain and also architects, engineers and teachers of drawing and fine arts who have carried out essential research work. In an orderly and coordinated manner, EDICES has developed the three planned phases of the study of the Sudarium:

- The first phase (1989–1994) consisted in a descriptive study of the cloth and a detailed analysis of the major stains found on the Sudarium.

- In the second phase (1994–2007), the medicolegal analysis of all stains and the elaboration of a comprehensive hypothesis about the use of the Sudarium were concluded.

- The third phase, which began in 2007, is performing a multi-disciplinary study on the possible identification between the *Man of the Sudarium* and the *Man of the Shroud*. ("Men" whose levels of coincidence are such that they exclude the chance and pointing to the fact that they are the same person).

The main results of EDICES' investigations were presented at two international specific congresses (1994 and 2007) sponsored by the University of Oviedo and the City Council of the capital, which published the proceedings of the congress sometime later.

Logically, the third phase of the investigation is the one that has greater importance and significance for the media, but it could not have been approached without the immense previous work performed almost anonymously by the different components of the research team for whom we will never have enough words to thank them for their work—it is noteworthy that none of the members looks for personal prominence.

In addition to the bibliography of the publications of EDICES members included in this book, there are the exhaustive articles in Spanish presented in the CES journal *Linteum*. For example:

- AA.VV. *"¿El Sudario de Oviedo y la Sábana Santa de Turín, dos reliquias complementarias?"*, Línteum n° 4. (December 1990) 1991 Valencia 1991.
- VILLALAÍN BLANCO, J. DELFÍN. *"Estudio hematológico forense sobre el "Santo Sudario" de Oviedo"*, Línteum n° 12–13 (October–December 1994) Valencia 1995. pp. 5–11 and 21–24.
- RODRÍGUEZ ALMENAR, JORGE MANUEL. *"El Sudario de Oviedo: Cómo se usó"*, Línteum n° 24–25. (December 98–March 99) 1999 Valencia. pp. 10–14.
- HERAS MORENO, GUILLERMO. *"II Congreso Internacional sobre el Sudario de Oviedo: Conclusiones"*. en Línteum n° 42–43 (January–December 2007) Valencia 2008. pp. 26–37.
- MIÑARRO LÓPEZ, JUAN MANUEL. *"Sobre la compatibilidad de la Síndone y el Sudario"*, Línteum n° 56 (January–March 2014) Valencia 2014. Monographic issue.

The members of EDICES also presented papers to the congresses on the Shroud of Turin held in the Republic of San Marino (1996), Valencia (2012), Bari (2014), St. Louis (2014) and Pasco (2017).

In addition, EDICES has produced complementary reports for the Oviedo Cathedral Chapter and made some partial publication of such studies. Some of its members have also participated personally in the congresses on the Shroud of Turin, explaining the compatibility of both cloths.

It should be added that the impact of the conducted research has made it possible for currently four members of EDICES to form part of the international scientific team of the *International Center for Studies on the Shroud* in Turin.

C. César Barta's Book

This present work, at least, has the great merit of being the first book in English on the research conducted for more than thirty-five years by the CES on the Holy Sudarium of the Cathedral of Oviedo. During this time, we have read some research written by "scholars" of which we are aware that they had not even seen the Sudarium and did not know about our research in depth, and we realized how easy it is to interpret improperly (or simply out of context) the data. Fortunately, this book will provide the opportunity to learn first-hand—through the pen of someone who belonged to the research team for years—about many data which are published for the first time in the language of Shakespeare, transcending the scope of the Spanish language.

I hope that this work will contribute decisively to the knowledge of the Sudarium. So, I wish it every success, because it will make our work and data taken directly on the relic accessible to the Anglo-Saxon world. That is already a very important contribution, and this book should be a point of reference for scholars of the subject.

Finally, I want to highlight that César Barta does not limit himself to copying or picking up what the CES research team (EDICES) has said or published, but rather he contributes with his own personal conclusions, with which we may agree or not, but that is something allowed and perfectly acceptable as long as the data are respected, which is manifestly done here.

Note

This book belongs to the Jenny Stanford Series on Christian Relics and Phenomena. Although it is focused on the Sudarium of Oviedo, there are some mentions of and relationships to other objects assumed to be relics included in other books of this series. The approach of different authors is not always the same and some discrepancies among different books should be expected. The research in this field, as in any other scientific one, allows working with different hypotheses to explain the same observation and it is not always possible to perform an experiment to discriminate. The reader has to evaluate between the reasons provided among the books in this series and retain the most persuasive.

Chapter 1

Introduction

In this chapter, a detailed description of the Sudarium of Oviedo is presented with its summary of use. We refer this cloth and other alleged sudariums to the Gospel of St John 20:7. Finally, we provide some perhaps unnoticed information about other interesting relics of Christ.

1.1 Description of the Sudarium: Size and Type of the Textile

There are many relics that claim to have come from Jesus Christ—perhaps too many. An analysis is necessary to determine if a relic is genuine or not[1]. In the case of the relics of Jesus Christ, a conclusive analysis is not always possible. Hopefully, there are some exceptions. Among the relics that contain elements that allow scientific research is the Sudarium. It is one of the relics that allows the analysis of its features. In this book, the experiments and studies that lead to a conclusion will be presented. However, to facilitate understanding the following chapters, it is very useful to present some conclusions first.

_____

[1]There are also secondary relics obtained by contact to genuine relics.

_____

The Sudarium of Oviedo: Signs of Jesus Christ's Death
César Barta
Copyright © 2022 Jenny Stanford Publishing Pte. Ltd.
ISBN 978-981-4968-13-3 (Hardcover), 978-1-003-27752-1 (eBook)
www.jennystanford.com

Most of the scientific studies performed on the Sudarium of Oviedo (Sudarium in this book) have been carried out by the Research Team of the Spanish Center of Sindonology (EDICES from the Spanish name *Centro Español de Sindonología*). The story behind this group and its activities will be described later. At this point, it is enough to say that the contribution of these scientific studies to the knowledge of Christ's burial has been conclusive enough.

The Sudarium that is being preserved at the cathedral in the town of Oviedo, located in the north of Spain, is made of linen with dimensions about 83 cm × 53 cm. Fig. 1.1 shows a photograph of the Sudarium taken in 1989; on the right is Dr. Jorge Manuel Rodriguez Almenar, current president of the *Centro Español de Sindonología* (CES), and Dr. Teresa Ramos Almazar, professor of Legal Medicine, during the first observation of the Sudarium granted to the EDICES.

Figure 1.1 Jorge Manuel Rodríguez and Teresa Ramos measuring the Sudarium of Oviedo (© CES).

The photo gives the reader a first impression of the size of the Sudarium in comparison with a human. The members of the team are measuring the cloth in its old reliquary. The Sudarium is on the Holy Ark (chest) and they are in the *Cámara Santa* (Holy Chamber) at the Oviedo Cathedral. The layout of this place has changed now, as we will see later.

Let us now focus on the description of the Sudarium. The CES team made a geometrical study of this linen and proposed a reconstitution of the use of the Sudarium[2]. The most conclusive result of the forensic research is that the Sudarium was used around the head of a person who died in vertical position (Fig. 1.2).

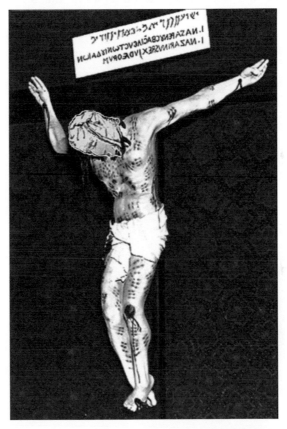

Figure 1.2 Use of the Sudarium around the head of a corpse (© CES). [*Note*: The purpose of this image is only for showing the use of the Sudarium. It does not claim to represent the actual posture of the body on the cross. This note applies to every instance of this type of image.]

---

[2]Guscin, Mark. (1997). Le Soudarion d'Oviedo: Son histoire et ses liens avec le Linceul de Turin. RILT, **4**. pp. 2–9.

In this first phase when the Sudarium was applied on the corpse, the cloth wrapped the head of a man who remained in a vertical position after having died approximately an hour before. The Sudarium had one of its edges on the nape of the corpse. The rest of the Sudarium passed over the left ear of the corpse to cover the face. It did not reach the right ear and left it uncovered (Fig. 1.3 and Fig. 1.4). Near the right ear, the cloth was folded again to turn towards the face creating a second layer in front of the mouth and nose area. To understand the stains seen on the Sudarium, it is very important to note that there were two layers that collected the fluids that came out from the mouth and the nose.

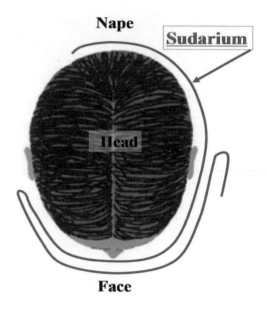

Up sight

Figure 1.3 Arrangement of the Sudarium on the head. Top view.

With some frames extracted from a video demonstration performed with a volunteer (Fig. 1.5), we try to explain how the Sudarium was used with the corpse still on the cross.

In this mode, the part of the Sudarium over the nape has received small stains of blood coming from wounds produced by objects with a sharp point. The part that covered the face received a mixture of blood and pulmonary edema that ran from the mouth and the nose: The corpse had both a beard and moustache.

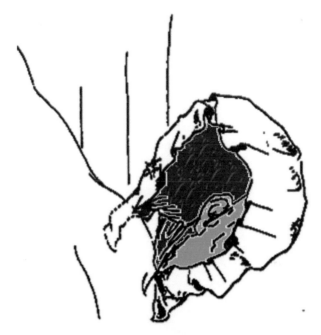

Figure 1.4 Arrangement of the Sudarium on the head. Right ear view.

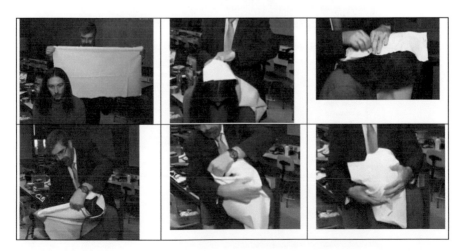

Figure 1.5 Frames from a video demonstration to explain how the Sudarium was used.

If the Sudarium of Oviedo was used for Jesus Christ, it was placed when He was still on the cross and it stayed until his arrival in the tomb. Just before the corpse was covered with the shroud, the Sudarium was removed and put aside close to the corpse in the sepulchre.

Now that we have a view of the use, we can show the cloth in some detail. Figure 1.6 shows the side that was in contact with the head. Near the left edge, at the lower part, we notice the stains that came from the nape. We can see, almost in the lower corner, a stain called the butterfly stain that was over the lock of hair falling on the upper back of the corpse. A little up, we see the small stains of blood that match with wounds of a crown of thorns.

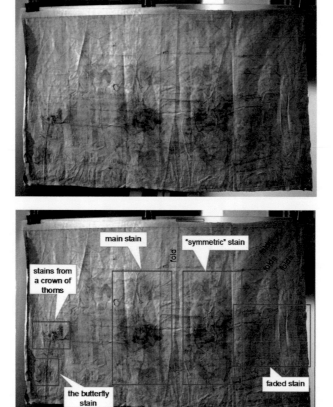

Figure 1.6 Sudarium out of reliquary (© CES) and location of main features.

If we move our sight toward the middle, we pass over an area with a very faded stain to reach at the main stain near the middle. The lower part of this main stain was in front of the mouth and the nose. The upper part of this stain was in front of the forehead. The stain of the mouth is darker than the stain of the forehead.

If we look at the right side, we find a stain "symmetric" to the previous one. Between these two main stains, there is a fold that runs from top to bottom splitting in two parts near the lower edge. The cloth was folded by this fold when the two main stains soaked the Sudarium. This explains their symmetry. Moving to the right we find another faded stain that arrives to the edge on the right.

Apart from the fold associated to the symmetry; there are other folds that evidence how the cloth has been kept for centuries. Moreover, a group of folds at the upper corner at the right in an oblique direction point to the evidence of an ancient knot.

Furthermore, the Sudarium presents some contaminations that are not related with the primordial use of the linen: fungal and spore colonies, pollen, presence of wax drops, burns, fingerprints of carmine, traces of silver paint, remains of adhesives, environmental contamination, tears, etc.[3] Their analysis will be the subject of later chapters.

Now let me describe a contamination that is perhaps the most astonishing of them. It is the remains of silver paint. On the right half of the cloth in the Fig. 1.6, between the symmetric stain and the right edge there are some lines and square angles that correspond to the outline of two rectangles of about 185 × 100 mm. The most realistic hypothesis of their origin is the contact with the bottom of a can which contained silver paint. We can reconstruct the fingerprint of the two simulated positions of the can on the cloth (Fig. 1.7). In this image there is also a visual

---

[3]Montero, F. (2007). *Descripción Química y Microscópica del Lienzo*. Oviedo relicario de la Cristiandad. Proc. II International Congress on The Sudarium of Oviedo, Oviedo. pp. 103–124. Also, Montero, F. (1994) *Descripción del Lienzo Química y Microscópica*. "Sudario del Señor", Proc. I International Congress on The Sudarium of Oviedo, 1994. pp. 67–82.

magnification of one of the silver stains corners. The threads are crushed under the weight of the can. The silver nature of the stain was confirmed by an Energy Dispersive X-ray microprobe[4]. We can assume that a worker was commissioned to repair the frame of the reliquary of the Sudarium and he did not care about the cloth. This event perfectly shows the attitude of some relic custodians. In conversation with ecclesiastics during my research, I perceived, although rarely, that some of them were more concerned about the container than about the content. They were more involved with the artistic value of the reliquary than with the religious value of the relic itself.

There is also a superficial contamination of blackish micro particles containing silver. They are spread along the top of the threads and of the wrinkles. The origin of this could be from the ancient rubbing with a silvered reliquary.

To have the most detailed description of the Sudarium of Oviedo, the reader can read through the following chapters and also look at Montero (2007)[5].

Figure 1.7 Silver contamination location.

---

[4]Montero, F. (1994) *Descripción del Lienzo Química y Microscópica.* "Sudario del Señor", Proc. I International Congress on The Sudarium of Oviedo, 1994. pp. 67–82.
[5]Montero, F. (2012). *Description of the Sudarium of Oviiedo.* 1st International Congress on the Holy Shroud. Valencia 28-30 de April 2012. Montero, F. (2007). *Descripción Química y Microscópica del Lienzo.* Oviedo relicario de la Cristiandad. Proc. II International Congress on The Sudarium of Oviedo, Oviedo. pp. 103–124.

1.2   Gospel of St John 20:7

Let analyse the passage about the sudarium in the Gospel of John 20:6-7. It reads in different translations to English as following:

> **New American Bible**. When Simon Peter arrived after him, he went into the tomb and saw the <u>burial cloths</u> there, and the <u>cloth that had covered his head</u>, not with the burial cloths but rolled up in a separate place[6].

> **International Standard Version**. At this point Simon Peter arrived, following him, and went straight into the tomb. He observed that the <u>linen cloths</u> were lying there, and that the <u>handkerchief that had been on Jesus' head</u> was not lying with the linen cloths but was rolled up in a separate place.

> **English Standard Version**. Then Simon Peter came, following him, and went into the tomb. He saw the <u>linen cloths</u> lying there, and the <u>face cloth, which had been on Jesus' head</u>, not lying with the linen cloths but folded up in a place by itself.

> **American Standard Version**. Simon Peter therefore also cometh, following him, and entered into the tomb; and he beholdeth the <u>linen cloths</u> lying, and the <u>napkin, that was upon his head</u>, not lying with the linen cloths, but rolled up in a place by itself.

Now, let us discuss what the Sudarium mentioned by John can be. The passage of John 20:7 speaks about the Sudarium (σουδαριον) which was placed on the head of Jesus. It has been supposed by some authors to be either a headdress or a chin band[7]. There were even authors who identified this piece of cloth with the Shroud of Turin itself. In the First International Congress on the Sudarium of Oviedo[8], the Canon Enrique López Fernández, professor of Latin and Holy Scripture, pointed out the opinion of some other scholars proposing that the word sudarium is a transcription of a Hebrew or Aramaic word *sûdar*

---

[6]New American Bible. http://www.vatican.va/archive/ENG0839/PXS.HTM. Original Greek *καί τό σουδάριον, ὄῆν ἐπί τῆς κεφαλῆς αύτο, οὐ μετὰ τῶν ὀθονίων κείμενον ἀλλά χωρίς ἐντετυλιγμένον είς ενα τόπον.*

[7]Antonacci, M. (2015). *Test the Shroud*. (LE Press. USA). Appendix G. pp. 358–359.

[8]AA. VV. (1994). *Sudario del Señor*. Proceeding of the First International Congress on The Sudarium of Oviedo. University of Oviedo. Oviedo.

or *sûdara*. These last can mean a cloth covering the entire body. As we will see later, the Sudarium must be a cloth of much reduced size than the Shroud. Leaving aside this last interpretation, which has already been given up by most authors, we are going to analyse the other versions corresponding with a linen of smaller dimension[9].

The headdress or hair band would be the "Saint Coiffe" of Cahors (France) which was used to cover a head of a man like a cap. The chin band would be a strip that goes below the chin and over the ears to reach the top to keep the mouth closed. Nobody has claimed to keep it preserved. In this book, we propose that the sudarium of the Gospel of Saint John should be the Sudarium of Oviedo in Spain.

We will analyse the subject from different points of view starting with the linguistic discussion.

1.2.1 Word Sudarium

According to Guscin [1999][10], only John's gospel uses the Greek term τό σουδάριον in relation to the burial of Jesus. The Greek word is actually a loan word from the Latin **sudarium**, which is a difficult word to translate into English. The various versions of the Bible have used such terms as "napkin", "face cloth" and just "cloth". Its etymology, closely linked to the Latin **sudor** (sweat), strongly suggests it was used to wipe sweat from the face and clean it, in general.

A 5th century text remarks the singularity of the word *sudarium*: Nonnus of Panopolis[11], an Egyptian writer, when paraphrasing the Gospel of John, Chapter 20, added this comment:

> Peter saw the linens there together and the cloth (ζωστηρα) that covered his head, with a knot toward the upper back of the part that covered the hair. In the native language of Syria, it is called sudarium. It was not with the funerary linens, but was rolled up, twisted in a separate place.

---

[9]Barta, C., and Guscin, M. (2004). *What is the Sudarion?*. *RILT* **26**. pp. 22–33.

[10]Guscin, M. (1999). *Recent Historical Investigation on the Sudarium of Oviedo*. Proc. Shroud of Turin International Research Conference. Richmound, VA. Glen Allen. VA: Magisterium Press.

[11]Nonnus of Panopolis (Egypt) writer in 400–479 A.D. Bennett, Janice. (2001). *Sacred Blood. Sacred Image. The Sudarium of Oviedo*. (Libri de Hispania). Colorado. p. 22.

Now, it is only to be noted that Nonnus had to explain what the cloth was, because he realized that he did not have a proper translation for sudarium. And another point to note is that the sudarium was not placed with the Shroud but in a separate place.

The word "sudarium" is also mentioned three more times in the New Testament and the context allows us to recognize the kind of linen that this word denotes[12]. In Luke 19:20 the sudarium is used to hide a small piece of money:

> Luke 19:20. And another came, saying, Lord, behold, here is thy pound, which I have kept laid up in a napkin.

Sudarium is also mentioned in the Acts 19:12, The sudarium that had touched the body of St Paul is utilized to heal the sick.

> Acts 19:12. So that from his body were brought unto the sick *handkerchiefs* or aprons, and the diseases departed from them, and the evil spirits went out of them.

In John 11:44, the sudarium is utilized for the death of Lazarus, in a very similar way of what you find in John 20:7.

> John 11:44: And he that was dead came forth, bound hand and foot with grave clothes: and his face was bound about with a napkin. Jesus saith unto them, loose him, and let him go.

A study of the word in classical sources[13] (Catullus, Petronius, Suetonius, and Martial) gives the idea of a cloth equivalent of a large modern handkerchief, or napkin or towel. The size of the Sudarium excludes its identification with the Shroud of Turin, as we have already said. Moreover, it seems that this piece of cloth can be used not only for funeral rituals, but it could have other general uses. It was normally carried round the neck or tied to the wrist or, even as a turban[14]. This suits well with the four texts of the New Testament and with several Latin texts[15].

---

[12]King James Version of the Holy Bible (1769) public domain.

[13]Guscin, M. (1999). *Recent Historical Investigation on the Sudarium of Oviedo*. Proc. Shroud of Turin International Research Conference. Richmound, VA. Glen Allen. VA: Magisterium Press.

[14]O'Callaghan, J., cited by Bennett, Janice. (2001). *Sacred Blood. Sacred Image. The Sudarium of Oviedo*. Libri de Hispania. Colorado. pp. 147–148.

[15]Guscin, Mark. (1998). *La posición de los lienzos en el sepulcro*. Del Gólgota al Sepulcro. Centro Español de Sindonología. Valencia. pp. 131–133.

1.2.2 On the Face

To look into the way the Sudarium was used, we refer to the discussion of Barta [2004][16]. The Gospel says that the Sudarium had covered the head. Here we note that a cloth that only covers the head could preclude a description saying that it covers the face. However, a cloth that covers the whole head including the nape, the ears, the cheeks, and the face, allows a description saying that it covers the face. This point introduces the following paragraphs.

The text of John 11:44 talks about the sudarium of Lazarus. It suggests thinking about linen that Lazarus wore <u>on</u> his face. All the consulted translations in several languages give "face" as translation of οψις[17]. The text of St John says:

> The man who had died came out, his hands and feet bound with linen strips, and his face (οψις) wrapped with a cloth[18] (σουδαρίω sudarium).

John does not give us a detailed description of all the garments of the deceased, and only mentions the linens that represented a difficulty for the deceased to unravel by himself. This context makes us think that the sudarium implicated a difficulty. This could be because the sudarium, covering the face, prevented from seeing well. The utilization of a sudarium that wrapped the head in order to hide the face is well compatible with the most natural translation. If Lazarus had worn a cloth that did not prevent his seeing, then we would have had to find other reasons to mention this part of the garments.

Besides the Gospels, there are two ancient texts speaking about the sudarium of John 20:7. The gospel of Nicodemus or Acts of Pilatus and the account of a pilgrimage of devotees of St Anthony of Piacenza. The redaction of the first is finished in the 5th century and the second is of the 6th century.

---

[16]Barta, C., and Guscin, M. (2004). What is the Sudarion?. *RILT*. **26**. pp. 22–33.

[17]King James Version, International standard Version, English standard Version, American standard Version, etc. The original Greek is: ἐξῆλθεν ὁ τεθνηκώς δεδεμένος τούς πόδας καί τάς χειρας κειρίαις, καί ἡ ὄψις αυτού σουδαρίω περιεδέδετο. λέγει αὐτοῖς ὁ ἰησουός· λύσατε αὐτόν καί φετε ὑπάγειν.

[18]English Standard Version.

The gospel of Nicodemus in the 15th chapter, paragraph 6 says:

"I am Jesus, whose body thou hast requested from Pilate and thou *hast* clothed me with a clean shroud, and thou hast put a <u>sudarium on my face</u> (προσωπον)"[19].

and a little further the same text repeats:

"He (Jesus) showed me the place where the shroud *was lying and the sudarium that had been used for his face* (προσωπον)"[20].

These paragraphs of the rather ancient cycle of Pilatus use the word "*sindon*" instead of "*othonia*" and make a clear distinction between *sindon* and *sudarium*. We also stress the use of the word "face" instead of "head".

The other ancient text tells us about the trip of a group of Italians to Palestine, and when they went past the Jordan River, they found there a grotto that lodged seven virgins:

"The sudarium which was on the face (fronte) *of our Lord is said to be there"*[21].

Again, the use of the word "face" instead of "head" is to be stressed. Therefore, in the 5th and the 6th centuries, the writers thought that the sudarium of the sepulchre could cover the face.

1.2.3 Aside

Finally, let us deal with the place of the sudarium in the tomb. As Guscin says[22], Saint John underscores three times that the Sudarium was actually separated from the other cloth: "*not with the other cloths*", "*separately*" and "*in a place by itself*". Some authors, however, have tried to read it as that the sudarium was inside the Shroud at the place that was the head of Christ. Let me clarify here that this interpretation is not issued from the linguistic

---

[19]καί ενεδυσας με σινδονι καθαρα και σουδαριον επεθηκας επι το προσωπον μου.

[20]καί το σινδονιον εκειτο εν αυτο και το σουδαριον το εις το προσωπον αυτου.

[21]in ipso loco dicitur esse sudarium qui fuit in fronte Domini.

[22]Guscin, M. (1999). *Recent Historical Investigation on the Sudarium of Oviedo.* Proc. Shroud of Turin International Research Conference. Richmound, VA. Glen Allen. VA: Magisterium Press.

analysis of the text, but because those authors assume that the sudarium was a chin band that was around the head of Christ when He was inside the Shroud. When the body disappeared, the chin band remained in its place inside the Shroud. The same phenomenon would have happened if the Sudarium of the Gospel of St John 20 was the "*sainte coiffe*".

After establishing these three features, meaning of the word sudarium, its possibility to cover the face and its being placed aside of the Shroud, we shall evaluate which option can better fit as candidate to be identified with the Sudarium of the St John's Gospel.

1.2.4 The Sainte Coiffe

The **Sainte Coiffe** or headdress (Fig. 1.8) is preserved in the St James cathedral of Cahors. It is a piece of cloth that has the dimensions of a hair band which has been made to cover a head of a man like a cap and it runs down to be buttoned under the chin[23].

Figure 1.8 Sainte Coiffe or headdress of Cahors.

---

[23]Babinet, Nathalie et Robert. *Le Suaire de Cahors ou la "Sainte-Coiffe"*. RILT. n. 7 pp. 21–27.

It is composed of eight superposed pieces of linen. The opening of the headdress around the face has got a contour of a bit less than 60 cm[24]. On the part corresponding behind the head, a straight sewing combines both sections of cloth that cover the ear and the cheek of both sides of the face. Robert Babinet wrote articles in several issues of the Revue Internationale du Linceul de Turin (RILT)[25] claiming the identity of the sudarium of the Gospel of St John to be same as the Sainte Coiffe. He reported that this cloth was part of the Jewish funeral customs specially made to be utilized on the head. It is evident that the objective of this kind of dress is to be worn while the corpse was inside the Shroud. In this case, because the Shroud has the image of the head of the Man, if He had the Sainte Coiffe around his head, the Sainte Coiffe had to be affected by the image process that branded the Shroud. It is not the case.

The authors describe the headdress as leaving only the face exposed, from the middle of the forehead to the chin. This arrangement is already incompatible with the image of the Shroud. In order to allow the discussion, it should be assumed that the head band was placed to allow seeing the entire forehead and the part of the hair which are observed on the image of the Shroud of Turin.

We find that this kind of headdress cannot be described using the same word (sudarium) that was used to designate the handkerchief in Luke 19:20 and that was used to save a coin. And, if the purpose of the saint headdress was to be used for deceased, it makes no sense to describe it as the cloth that Saint Paul wore when alive in the Acts 19:12.

It does not seem to be the kind of sudarium of Lazarus who wore it on his face. The saint headdress only covers the head. No version of this verse has been translated by "*pathil*" that, according to Babinet, is the Jewish word for this burial cloth. The *pathil* was a small fabric that surrounded the hair to keep it together. The *pathil* was tied under the chin to keep the mouth closed[26]. However, according to Nonnus of Panopolis the

---

[24]https://fr.wikipedia.org/wiki/Sainte_Coiffe_de_Cahors#cite_ref-22.

[25]Babinet, Nathalie et Robert. *Le Suaire de Cahors ou la "Sainte-Coiffe"*. RILT. n.7. p. 23.

[26]Climont, Jean de. (2016). *The Mysteries of the Shroud*. Assailly Editions, Paris 2016. p. 43.

sudarium had "*a knot toward the <u>upper rear</u> that covered the hair*". The detail of the knot signalises a supplemental source of information of the gospel that disagrees with the identification of the Saint Headdress (*pathil* or *coffia*) because the knot in this garment is not in the upper part of the head but under the chin.

If we analyse the case of Lazarus, he could have worn a *pathil* but it was not the cloth St John spoke of when he mentions the sudarium. It is confusing to describe a hair band as going around the face. For a hair band, it is more natural to say: "it was on the head".

The geometrical analysis provides strong evidence against the possibility that the *sainte coiffe* had been over the corpse head when the image of the Shroud was impressed. Even if measurements could not be made on the site of Cahors, it was possible to draw some rather precise estimation from a photo of the *sainte coiffe* shared by RILT N. 7, p. 21. The reconstitution of the measures has been effectuated by means of the contour of the edge, i.e., 60 cm, as the authors of the article indicated[27]. This measure exactly corresponds with the outline of the head of a middle-aged person. The length of the outline gives us the scale of the photo. The reliquary, having a circular base, allows us to estimate the depth by means of the length. So, it is possible to reconstruct a three-dimensional model.

One of the most interesting measures taken there is the length from the forehead to the nape: 27 cm. In the article quoted above, the authors justify the presence of the saint headdress on the head of the Man of the Shroud because the lacking image between the frontal and dorsal figures of the Shroud. However, we find a different conclusion when we estimate the measures. Indeed, the separation shown by the Shroud between the two images of the head is less than 19 cm. Thus, there would be more than 8 cm hiding the back of the head of the Man of the Shroud, if he had worn the saint headdress of Cahors.

The superposition of the Sainte Coiffe over the dorsal image of the head of the Man of the Shroud is shown in the Fig. 1.9.

---

[27]Babinet, Nathalie et Robert. *Le Suaire de Cahors ou la "Sainte-Coiffe".* RILT. n. 7. p. 21.

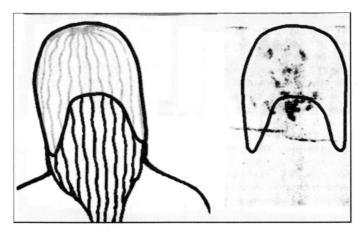

Figure 1.9 *Sainte-Coiffe* of Cahors superimposed on the head of the Man of the Shroud.

If the saint headdress had been on the head of the Man of the Shroud, it would have prevented from the formation of the image on a large part of the head on the dorsal side, which is however easily visible in the positive as well as in the negative image of the Shroud. Also, the stains of blood would not be clear, because the saint headdress would have absorbed most of the blood. Moreover, such stains are absent from the headdress of Cahors. Thus, there is only one possible conclusion: the *sainte coiffe* of Cahors was not put around the head of the Man of the Shroud when the image was impressed.

The lack of image between the frontal and dorsal image of the Shroud used as support of the presence of the Saint Headdress is weak because this absence of image between frontal and dorsal part of the head of the Man has been explained by the vertical projection characteristic of this image. There is no image between frontal and dorsal part of the head as there is no image on the sides of the Man. The upper part of the head follows an approximately vertical orientation like the elbows or the outer parts of the legs, neither of which left an image.

Those authors, who support the thesis of a hair band or a chin band, imagine quite logically that it was around the face of the corpse and, then, it would have remained between the two halves of the Shroud when the corpse disappeared. This is the logical scenario, but it cannot be the sudarium of the verse of the gospel that was clearly set aside the Shroud. The chin band scenario will be discussed in the following section.

Finally, it is astonishing to imagine a Jewish corpse buried dressed in a hat only and without any other cloth. We must conclude that the *Sainte Coiffe* of Cahors is not the Sudarium of the Gospel and it has not been utilized for the Man of the Shroud of Turin when he was inside it.

In the year 2019 the Holy Headdress was venerated and carried in a procession through the streets of Cahors. The Diocese of Cahors celebrated the 900th jubilee of its Saint-Etienne Cathedral. Despite the above conclusion, there is nothing to oppose the wishes of Pope Francisco pronounced for this occasion:

> That the relic of the Holy Headdress helps Christians, as well as all men of good will, to open their hearts to the Lord Jesus who is come for all to have life in abundance.

1.2.5 The Chin Strap

The **chin strap** is another candidate to be equated to the Sudarium of the gospel of St John. The hypothesis of the presence of the chin band has been suggested to explain how the locks of hair that flank the face the image of the Man of the Shroud are arranged. These locks of hair run along the cheeks to almost reach the shoulders and make suppose a support[28]. If the chin band were fit with a knot at the top of the head, it could also explain the lack of the image on this area. In addition, some authors have found instructions in the Jewish *mishnah* commanding to attach the chin of a corpse in order to keep his mouth closed[29]. The researchers of the STURP (Shroud of Turin Research Project), have simulated the presence of a chin band that goes from the top of the head, over the ears to reach below the chin (Fig. 1.10). In this simulation, the knot was not on the top of the head but under the chin. It does not correspond to the knot mentioned by Nonnus of Panopolis.

---

[28]Gail (de), Paul. (1970) *Le visage de Jesus-Christ et Son Linceul.* France Empire. p. 260.

[29]Steveson K. Et Habermas G. (1981). *Veredict on the Shroud.* Michigan: Servant Press. Spanish translation: *Dictamen sobre la Sâbana de Cristo*, Planeta 1988. p. 64.

In the case of the chin band, the cloth could have been a sudarium without a specific purpose merely as something worn around the neck or tied to the wrist and used to tie the jaw of Jesus in an unforeseen circumstance. It may also be compatible with the information provided by Nonnus of Panopolis about the knot on the top of the head. However, it is quite improbable that a chin band were the kind of sudarium of Lazarus.

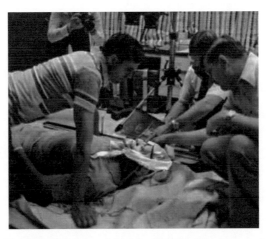

Figure 1.10 Chin Band simulated by the STURP (The National Geographic Magazine June 1980. p. 749).

A cloth the purpose of which was to keep the mouth closed indicates an impediment and justifies its mention among the garments bothering the freedom of Lazarus. However, as it was said for the saint headdress, a chin band was not a cloth that would have been described as been worn in front of the face. This is a subtle nuance. For a chin band, it is natural to describe it as going *around* the face. However, it would never be acceptable to describe it as being covering the eyes. In fact, the Greek word ὄψις used in the Gospel text is better used for sight[30], eyesight, vision, or glance than the face. Moreover, some writers propose to understand that the sudarium of Lazarus hid him from the sight of the others[31].

---

[30]E-Sword (App). The sword of the Lord with an electronic edge. 1909 Spanish Reina Varela (SRV+). John 11:44.

[31]López Fernández, E. (2004). *El Santo Sudario de Oviedo*. MADU. Granda-Siero. p. 50.

However, the difficulty to identify the Johannine *soudarion*[32] with a chin band comes from the place where the chin band must remain. López (2004)[33] argued that the sudarium was not a proper part of the shrouding, first at all, because the synoptic evangelists do not mention a sudarium being used during the burial. They only say that Christ was wrapped in a shroud (*sindon*). Moreover, if a chin band was used for the shrouding, it would have been placed beneath the shroud and Peter would not have seen it. The sudarium instead was separated from the Shroud.

To be sure of this conclusion, an experiment was performed during a session in the CES premises in 2015. A sculpture of the Man of the Shroud at half size was put inside a cloth simulating the Shroud also at half size (Fig. 1.11 and Fig. 1.12).

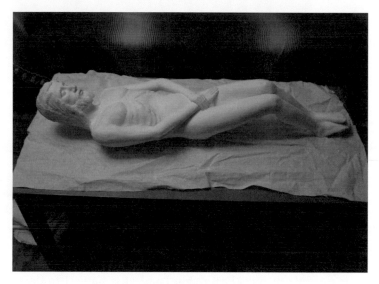

Figure 1.11 Man of the Shroud model (half scale).

There were about 40 people that took part in the unexpected experience. They were told that the arrival of the disciples Peter and John to the sepulchre in the early morning was going to be simulated and therefore the lights should be shaded (Fig. 1.13).

---

[32]καί το σουδαριον, ο ην επι της κεφαλης αυτου.

[33]López Fernández, E. (2004). *El Santo Sudario de Oviedo*. MADU. Granda-Siero. p. 51. Also referenced in Bennett, Janice. (2001). *Sacred Blood. Sacred Image The Sudarium of Oviedo*. Libri de Hispania. Colorado. p. 150.

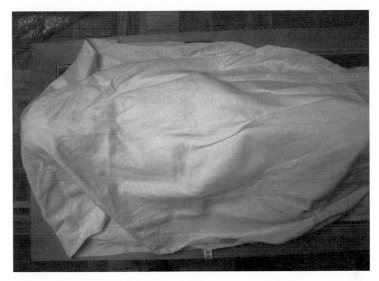

Figure 1.12 Shroud covering the model (half scale).

Figure 1.13 Sepulchre simulation with the model inside.

Then, almost in complete darkness, the sculpture was removed by two members and discreetly a chin band was hidden inside the shroud dummy and the shroud was allowed to fall freely. The result is shown in Fig. 1.14.

By chance, at the place where the head of the sculpture was and where the hidden chin band was placed, the cloth presented a small hill. In the following instants, people were asked to describe what they saw there. People only said that the cloths were creased and almost flat. A member noted the hill. But none of them mentioned the presence of the chin band.

Figure 1.14 Sepulchre simulation with the model removed.

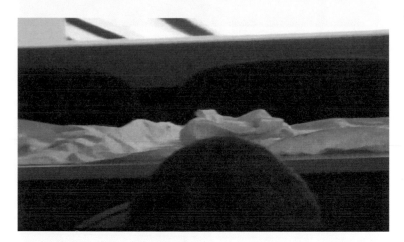

Figure 1.15 The chin band uncovered.

These were members of the CES, focused on the subject and having the text of the Gospel of John 20:7 available on a screen. But nobody could say anything about the cloth that was under the shroud because the chin band was not visible as it lay hidden between the layers of the shroud. If it could not be seen, even less could it be described — how was it rolled or flat or folded, etc. However, it was there but the sheet had to be pulled back to allow everybody to see it (Fig. 1.15). This experience showed that Peter could not see the chin band and could not say that it was folded or rolled up. The Gospel text is precise when it points out the position of the sudarium outside the shroud. The sudarium of the Gospel of John cannot be the chin band. The sudarium of the Gospel of John was something else.

We should accept that the chin band was not the Johannine sudarium. But the hypothesis of the presence of a chin band was justified mainly because three reasons:

- its knot explained the lack of image between images of the head in the Shroud,
- due to the need to close the mouth of the deceased and
- because the direction of the locks of hair.

Then, the chin band can be inferred from the other features although it was not the sudarium of the Gospel. The presence of a chin band among the linens used for the burial of the Man of the Shroud remains possible, but, as we will see, its presence was not necessary.

The lack of image between the two images of the head was justified above for the case of the *saint coiffe*. Moreover, there was no need to close the mouth of Christ. At the moment of death in vertical position, the head was falling in a natural way and the mouth could not have remained very open but must have become almost closed. The gospel reports the final inclination of the head of Jesus. Afterwards, the rigor mortis kept the mouth almost closed when Jesus was lowered from the cross, where He had remained with his head bowed on the chest[34]. After two hours of the inclination of the head, the mouth would have been almost closed definitively.

We still have the argument of the direction of the hair locks next to the cheeks when the corpse is being laid down on the back. In first place, the simulation performed by the STURP the chin band has the knot under the chin, it does not support those locks and crosses them at the level of the temples (Fig. 1.10). This last feature is not present in the image of the Shroud of Turin because the locks of hair start from the top of the head and run without any crushing up to the shoulders.

---

[34]Bennett, Janice. (2001). *Sacred Blood. Sacred Image. The Sudarium of Oviedo.* Libri de Hispania. Colorado. p. 150.

And, finally, the physical existence of the Sudarium of Oviedo and its utilization before the entombment of the Man of the Shroud, would explain by itself the position of the hair without the necessity of a chin band. We will see it in the following chapters of this book. The above-mentioned inclination of the head at the time of death of the crucified makes the locks fall forward over the cheeks (Fig. 1.16).

The blood leaking off the wounds produced by the crown of thorns mixed with sweat and dust would have given the required stiffness to the hair to keep them almost fixed. A sudarium around the head, like that of Oviedo, must have hardened the hair in its position, and after two hours with the head tilted and the hair tightened along the cheeks by the Sudarium, the hair would have taken a permanent shape, which is not easy to modify. The above thesis was verified with a practical simulation which is the subject of a later chapter in this book.

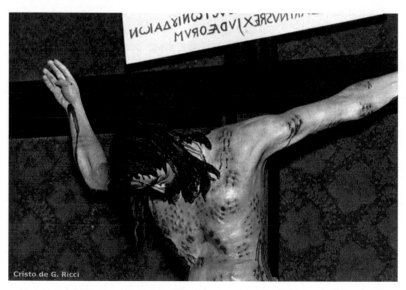

Figure 1.16 The sculpture promoted by Msgr Giulio Ricci. The locks of hair fall next to the cheeks (© CES).

We have seen that the Sainte Coiffe of Cahors and the chin band are not the sudarium mentioned by St John. The only candidate remaining is the Sudarium of Oviedo. We should remember that the conclusion of the study is that this Sudarium wrapped the head of a man who stayed in a vertical position at least one hour after his death. It was only after having been put in a vertical position with the Sudarium for about one more hour that the corpse was laid on the ground, face below, with the Sudarium still completely surrounding the head. Shortly after that, the man was moved and swayed. A few minutes later, the Sudarium was taken off and the cloth was embalmed with aloe.

Therefore, the Sudarium of Oviedo was used before the entombment while the corpse was on the cross *covering his head* and while he was lowered from the cross waiting for the permission to be buried. The first objective of its use was not to complete the garments of the burial customs but to collect the blood that came out from the nose and mouth after death and before burial. However, this cloth stained with the blood of the deceased also had to be buried with the corpse according to the Jewish traditions. We can assume that it was removed from the head before covering the deceased with the Shroud but it was put in such a way that it was *not lying with the linen cloths but was rolled up in a separate place.*

If the Sudarium of Oviedo was used for Christ, it was put around His head still on the cross and stayed till the arrival to the tomb. Just before the corpse was covered with the shroud, the Sudarium was taken off and put aside very close to the corpse. This fact is verified because the aloe applied for the shrouding left its traces until now in the stains of blood at the surface that was in contact with the face. It implies that the Sudarium had to be removed from the head to allow the aloe to reach this side.

1.2.6   The Sudarium on the Shroud

The **Sudarium on the Shroud** is another lesser known interpretation of the Gospel of St John (Fig. 1.17). According to some authors[35,36,37], the sudarium of the Gospel was outside the

---

[35]Persili, A. (1989). *Tutta la verità su Cristo Risorto.*Casa della stampa. Tivoli.

[36]Loconsole, M. (1999). Sulle tracce della sacra Sindone. Un itinerario storico-esegetico.Ladisa Editori. Bari. Italy.

[37]Tamiozzo, G. (2020). *Pregare con la Sindone – omaggio alla Passione di Cristo.* Tipografia Peretti. Quinto Vicentino. Vicenza. Italy.

Shroud but placed on top of it at the place of the head, covering this part of the Shroud.

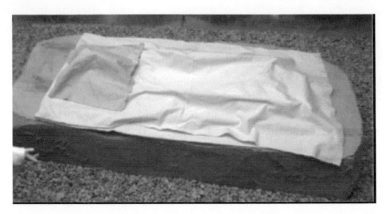

Figure 1.17 The sudarium, placed on the shroud, still remain in its place around the head even after the disappearance of the body (© Fanti).

This interpretation avoids the objection against the *sainte coiffe* and the chin band because this hypothesis leaves the cloth exposed. In fact, the scholar Don Antonio Persili described that the enveloping operation with the Shroud was preceded by the application of a sudarium inside the Shroud, where it acted as a chin guard, and followed by a second sudarium on the outside of the Shroud. These authors apparently searched an explanation of the act of faith of St John who *saw and believed* (John 20:8). What did the Apostle John see and in what did he believe? In particular, why did he believe? For them, the way the Shroud was laying was not enough to cause the act of faith. It was just the position of the sudarium that made John understand that Jesus had come out from those linens during His Resurrection. The hypothesis assumes that the sudarium was still wrapping the absent head, arranged in a particular way because it was stiffened by the oily substances in which it was previously soaked. The sudarium remained stiffened and raised in a curved form as if still enveloping Jesus' head due to the rapid drying of the salves. According to this hypothesis, John believed in the Resurrection only after having carefully observed the particular disposition of the sudarium when he entered the Sepulchre.

Although this verse of the fourth Gospel is often interpreted as the believing in the resurrection, Father St Augustine in the sermons 229 and 246 interprets that John thought that the corpse was stolen as the women had said. If St Augustine were right, no odd interpretation of the Gospel would be necessary and the whole argument would fall. However, against the authority of St Augustine we can admit that St John could believe in the resurrection but only secretly and in his privacy without sharing his faith with the rest of disciples, because the gospel of St Luke 24:24 clearly says that the disciples who went to the sepulchre did not believe in the resurrection but thought what the women said.

To justify the position of the sudarium on the Shroud, the followers of this hypothesis have proposed alternative translations from Greek. This is the Greek text:

Joh 20:7 καί τό σουδάριον, ὃ ἦν ἐπί τῆς κεφαλῆς αὐτοῦ, οὐ μετά τῶν ὀθονιων κείμενον, ἀλλά χωρίς ἐντετυλιγμένον εἰς ἓνα τόπον.

(kaí tó soudárion, ó ín epí tís kefalís aftoú, ou metá tón othoníon keímenon allá chorís entetyligménon eis éna tópon.)

One of the proposed translations for the above hypothesis is:

and the sudarium that was on his head not lying with the linens but rolled up separately in the same place.

This translation does not disagree with the four versions provided above (New American Bible, and International, English and American standards) except in the second last word: "same".

Don Lieto Massignani[38] states that the words "ἰς ἓνα τόπον" (eis éna tópon) correspond to a Semitism which indicates "*in the same place*". He considers that the translation "*in a place*" is not correct and that the translation "*in a separate place*" is even worse.

According to the proposed translation the sudarium would be in the same place. But what *same* place? In the *same* place that it was put? And where was it put? The text says that the sudarium was on his head. It is possible that the sudarium

---

[38]Personal communication through Giulio Fanti of Don Lieto Massignani, Professor of Sacred Scripture and Doctor of Medicine and Surgery, retired from facoltà Teologica dell'Italia Settentrionale-Padova. Italy.

was on his head during the burial. However, most of the translations say that the sudarium "*had been on Jesus' head*" and it can be understood as the sudarium was on the head of Jesus before the burial. This would be the case of the Sudarium of Oviedo.

However, with this single change in the translation, the position of the sudarium of the Gospel does not correspond yet to a cloth upon the Shroud because the Greek text says, "*not lying with the linens but rolled up separately*". The plain reading of these words leads to imagine that the sudarium was not with the Shroud. To support the hypothesis, we need more changes. The difference between the Shroud and the sudarium would not be the place but the way they stand.

For example, M. Loconsole[39] translates as "*and the sudarium, not stretched out like the bands, but wrapped differently in the same place*". This different arranging of the sudarium is justified because it was previously soaked by the oily substances which on drying caused the sudarium to remain stiffened and raised in a curved form as if still enveloping Jesus' head. This argument apparently assumes that the Shroud was not soaked with the same aromas and it did not remain as stiffened as the sudarium. If the body and/or the Shroud were also soaked, the way the Shroud remained after the disappearance of the body would be the same as the sudarium and even more accurately reproducing the shape of the face and the Shroud alone would be enough to cause the faith of St John.

Most of the authors who proposed these interpretations did not mention the Sudarium of Oviedo[40] and probably they were unaware of its existence. The ignorance of the features of the Sudarium of Oviedo justifies that they had to imagine other cloth to explain the Gospel. The Sudarium of Oviedo, as we will see throughout this book, had been on the head of the corpse and was removed from his head before the burial. The application of funerary spices on this cloth suggests that it was placed in the tomb separately from the other burial linens. It matches quite well the Gospel according to its standard translation.

---

[39]Loconsole, M. (1999). Sulle tracce della sacra Sindone - Un itinerario storico-esegetico - Ladisa Editori. Bari. Italy.

[40]Persili, A. (1998). *Sulle tracce del cristo risorto: con Pietro e Giovanni testimoni oculari.* C.P.R. This author analyses other identification of the sudarium but does not include a mention of the Sudarium of Oviedo. He should be unaware of its existence.

Another author (Fanti[41]) accepts the presence of the Sudarium of Oviedo in the sepulchre but he thinks there was another second sudarium which was most likely mentioned by St John. This other sudarium would be placed on the Shroud over the face place and the face was in turn covered by the Shroud. St John had seen what was used on Good Friday for the burial: The Shroud, the sudarium placed on the face perhaps lost over the centuries and the Sudarium of Oviedo. According to Fanti, the Sudarium of Oviedo was not allowed to be placed on the clean Shroud because it was impure as it was soaked in blood. If the proposed translation requires a sudarium on the Shroud, it would not be the Sudarium of Oviedo. Fanti is right: The Sudarium of Oviedo was removed from the head and very probably it was not put on the Shroud but put separately near the corpse.

According to most of the versions, the Gospel describes a sudarium (only one) separately from the Shroud. St John does not say "one of the sudariums" but "the sudarium". The Latin translation of the Vulgate under the authority of Jerome of Stridon confirms the modern usual translations:

> *et sudarium quod fuerat super caput eius non cum linteaminibus positum sed separatim involutum in unum locum*

> The sudarium... was not put [*positum*] with de linens [*linteaminibus*] but [*sed*] (placed) separately [*separatim*] in one [*in unum*] place [*locum*]

The strongest objection to the alternative interpretations is that the only translation of "οὐ **μεταὶ** τῶν ὀθονίων κείμενον" is "not lying **with** the linens". For the analyzed hypothesis, the text should be "not lying **like** the linens" or "not lying **as** the linens" using the Greek word "**ὥσπερ**", which does not appear in the verse. If there was another sudarium on the Shroud, it is just a speculation supported by an interpretation of the text about the position of the sudarium and the effect on the faith of John. It is a possible interpretation. However, according to the overwhelming majority of textual translation, John informs about only one sudarium and it was not on the Shroud. It was apart. Moreover, we recall the conclusion of Scholar Guscin that the

---

[41]Fanti, G., and Malfi, P. (2020). *The Shroud of Turin – First Century After Christ!*, Second Edition. Jenny Stanford Publishing Pte. Ltd. Singapore. pp. 41–43.

Gospel states three times that the sudarium was not with the Shroud. If there was another sudarium on the Shroud, it is not mentioned by John[42].

In conclusion, the *sainte coiffe* de Cahors was not rolled around the head of the Man of the Shroud of Turin when he was covered with the Shroud. A chin band is not necessary to explain the characteristics of the Shroud when one understands the utilization of a Sudarium like that of Oviedo. There is no need of a second sudarium on the Shroud to understand the Gospel. The Sudarium of Oviedo corresponds very well with the gospel of St John.

1.3 Other Relics of Christ

Before going deeper into the study of the Sudarium of Oviedo, it was thought advisable to dedicate some paragraphs to other relics. If not mentioned here, some personal stories might be forgotten forever.

Among the relics that claim to be from Jesus Christ there is one that stands out from the rest. The Shroud of Turin (the Shroud in this book) probably has been more investigated than any other archaeological object. A summary of the state of the research for the Shroud is that it is a true burial shroud of a man who was tortured with the same penalties as the Gospel describes for Jesus Christ, dated by carbon 14 in the 14th century and showing an astonishing image. There is a book in this series devoted to this Cloth[43]. We will briefly analyse the Shroud in this book. Moreover, there would be books on other relics in this series. In the meanwhile, other relics that deserve a mention are the *Title of the Cross* and the *Tunic of Argenteuil*.

---

[42]Fanti does not share this debatable interpretation letting the Author take the responsibility of his statements. According to Fanti, the argument would need deeper discussion elsewhere, also with biblical experts.

[43]Fanti, G., and Malfi, P. (2019). *The Shroud of Turin. First Century after Christ!*. Second Edition. Jenny Stanford Series on Christian Relics and Phenomena. Jenny Stanford Publishing Pte. Ltd. Singapore.

1.3.1 Title of the Cross

Michael Hesemann[44] carried out an important investigation about the Title of the Cross (see Fig. 1.18) and the Holy Cross that are in Roma in the Church of the *Santa Croce* (Holy Cross) of Jerusalem. There is probably the half of the Title with part of the three text lines of the Pilate sentence: Jesus of Nazareth King of the Jews.

Figure 1.18 *Titulus Crucis* preserved in the *Santa Croce* in Rome (© CES).

The paleography allowed dating the common time of the inscription styles in the first century. There are other characteristics that lead us to discard a falsification. A grammatical difference in the Greek text between the Gospel and the Title is in the adjective of the origin village, Nazareth[45]. In the Gospel the correct form appears as ΝΑΖΩΡΑΙΟΣ as it is expected for a canonical text. However, in the Title it is written as ΝΑΖΑΡΕΝΟΥΣ

---

[44]Górny, G., and Rosikon, J. (2014) *Testigos del Misterio. Investigaciones sobre las Reliquias de Cristo*. RIALP. Madrid. pp. 83–94. Translation of the original Polish Swiadkowie Tajemnicy.

[45]Hesemann, M. (2000). *La Reliquia del Titulus Crucis*. Linteum 27–28. December 1999–April 2000. p. 9.

that is just a transcription of the Latin form, understood if the writer is a worker that had to manage several languages. A forger would write just the canonical text. The other unexpected evidence of authenticity is that the Latin and Greek text are written from right to left following the direction of the Hebrew text of the upper line. While the conclusion for other relics only allows saying that they belong to somebody crucified like Christ or of the time of Christ, in the case of the Title, it can be qualified as the most undeniable relic of Jesus Christ because, once the fake is discarded, it is the only one that bears his name.

The fact that there is a relic of the Holy Cross in the same church associated to the Title, allows us to grant some credit to this particular piece. We can also tell that Pope Pious XII gave some credit to this relic saved in the Church of the Santa Croce of Jerusalem in Rome. This is the result of a story that happened in a village of Spain called Caravaca. In this place, located near Murcia, there was an assumed relic of the Holy Cross since centuries (see Fig. 1.19).

Figure 1.19 Caravaca cross.

During the Civil War of the last century in Spain, the relic was stolen, probably due to the economic value of the reliquary. However, the inhabitants had a yearly fest in honour of its relic with processions, solemn mass, etc. After the war, they needed their relic to maintain the celebration. They had pictures and detailed drawings of the reliquary and they commissioned a copy of the container. But it should not remain empty. On request from the bishop of the town, Pious XII sent a bit of wood taken from the relic of the church of the Santa Croce in Rome to be put inside the reliquary of the Caravaca[46].

1.3.2   Tunic of Argenteuil

Among the relics that bear elements allowing a scientific research there is also the Tunic of Argenteuil (see Fig. 1.20). It deserves a separate book in this series. There were two round tables to share the findings related to this relic. One took place in 1998 and the other in 2005. The main conclusions of the second round table were that the Tunic is a wool cloth with plenty of true blood with red cells. Some of them present a modified shape indicating a traumatic process[47].

Although the study of the blood group was not published due to the death of the researcher Saint Prix, the blood group was confirmed to be AB like the one found in the Shroud. The DNA indicated that the blood could belong to a Jew man[48].

I attended the first-round table of 1998. Professor André Marion showed his research about the Tunic. The proceedings of this first meeting were sent only to a limited group of people. The Marion's contribution was not included in the proceedings. However, some pictures of the screen were taken during the meeting. The infrared photography shows the main stains of blood (Fig. 1.21). Marion superimposed this picture to the outline

---

[46]Ballester Lorca, P. (2003). *La Cruz de Caravaca*. Caravaca, Murcia. Spain. p. 57.

[47]Lucotte, G. (2007). *La Sainte Tunique d'Argenteuil face à la Science*. Actes du Colloque du 12 Novembre 2005. Edited by Huguet, D. and Wuermeling, W. F-X de Guibert. p. 221.

[48]Jacquet, C. (2007). *La Sainte Tunique d'Argenteuil face à la Science*. Actes du Colloque du 12 Novembre 2005. Edited by Huguet, D. and Wuermeling, W. F-X de Guibert. p. 225.

of the Man of the Shroud (Fig. 1.22). Similar images can now be found in the relics book of Górny and Rosikon (2014)[49].

Figure 1.20 Holy Tunic of Argenteuil (© A. Marion).

It is worth mentioning that in this first round table, my presentation included the observation of the piece of the Tunic kept in the Cathedral of Toledo with the reputation of having its origin in the imperial treasure of Constantinople (Fig. 1.23). The thread of this tunic was also made of wool, color between brown and purple and spun Z (Fig. 1.24).

---

[49]Górny, G., and Rosikon, J. (2014). *Testigos del Misterio. Investigaciones sobre las Reliquias de Cristo*. RIALP. Madrid. p. 188. Translation of the original Polish Swiadkowie Tajemnicy.

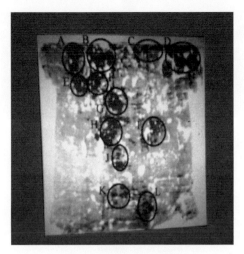

Figure 1.21 Tunic of Argenteuil. Infrared enhanced (© A. Marion).

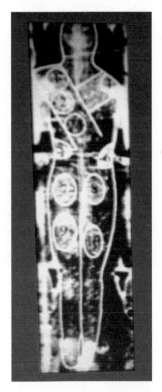

Figure 1.22 Tunic of Argenteuil. Infrared superimposed to the Shroud image (© A. Marion).

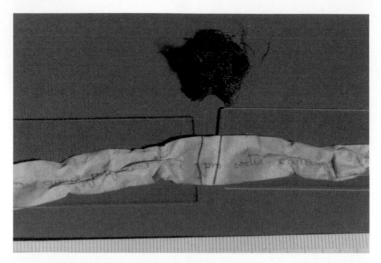

Figure 1.23 Part of the Tunic kept in Toledo.

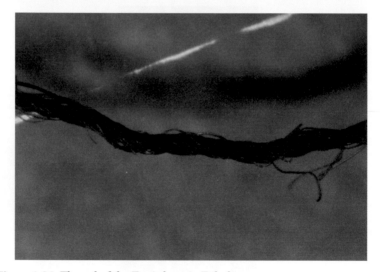

Figure 1.24 Thread of the Tunic kept in Toledo.

During the meeting, if I remember well, someone among the attendants stated that the spun of the Tunic of Argenteuil was S. In the proceedings, my conclusion was the impossibility that the tunic that was in the past in Constantinople was the same as that of Argenteuil. However, in my original draft of proceeding text there was a note, which disappeared in the delivered version, requiring a verification of the spinning of the Argenteuil Tunic. In the 2005 meeting, A. Marion reported the spun as Z[50] (as in the Toledo case) for a piece which is attributed to the Tunic of Argenteuil. Professor Giulio Fanti, having photos of yarns of the Tunic in his archive, also confirm the spun Z of the Argenteuil Tunic[51]. Now, it could be worth to verify the possible compatibility of both cloths. In the case that the conclusion is that two textiles are the same, a question would arise on the simultaneous presence of the Tunic in Argenteuil and one of its pieces in the Sainte Chapelle of Paris since 1250.

1.3.3   Sainte Chapelle of Paris

The story of the pieces of relics kept in the Toledo Cathedral[52] is one that needs to be told. Many Shroud investigators think that the Shroud of Turin was present in Constantinople from the middle of the 10th century until the beginning of the 13th century. The documents mentioned in the bibliography[53] enable us to arrive at quite a precise scenario of the time when Robert de Clari arrived in Constantinople with the IV Crusade.

The Imperial Palace in Constantinople was located in the South East part of the town. Inside this Imperial complex was the Pharos chapel. In the middle, there were two vessels of gold, hung on thick silver chains. One of the vessels contained a tile and the other one a cloth called the Mandylion. Both showed the face of Christ impressed on them. This was the Mandylion that was in Edessa and that was taken to Constantinople in 944, as is stated in the anonymous text attributed to Constantine VII *Porphyrogenetos*[54]. But De Clari's description of the legend associated with this icon does not mention either Edessa, or King

---

[50]Marion, A. (2007). *La Sainte Tunique d'Argenteuil face à la Science.* Actes du Colloque du 12 Novembre 2005. Edited by Huguet, D. and Wuermeling, W. F-X de Guibert. p. 213.

[51]Personal Communication.

[52]Barta, C. (2002). *The Louis IX Shroud Fragment.* Shroud Newsletter. N. 56, December. pp. 57 –65.

[53]Durand, J., et al. (2001). *Le trésor de la Sainte Chapelle.* Paris. Louvre. p. 50.

[54]Anonymous text *Story of the Image of Edessa.* Original in Greek. Translated into English by Wilson, I. (1978) *The Turin Shroud.* Book Club Associates. pp. 235–250.

Abgar or the time of Jesus Christ[55]. Another interesting church for the Shroud is Blachernae that was located at the other side of the town, to the North West. In this church, Robert de Clari saw the weekly exhibition of a Shroud with the figure of Christ raising itself upright. For a discussion about the incompatibility of the Mandylion and the Shroud see (Barta 2018)[56].

At that time, the Pharos chapel housed the most important collection of Christian relics[57]. We can enumerate: the cross, the nails, the lance, the sponge, the cane, the crown of thorns, the shroud of the tomb, the sandals, the towel with which He dried the apostles' feet, the tunic, the stone from the tomb and the sudarium[58]. In 1203, Robert de Clari[59] pointed out not only the presence of the Mandylion and many of the aforementioned relics, but he also mentioned separately two more relics not belonging to Christ: The Virgin's veil and John the Baptist's head.

We mainly focused on the Tunic, as mentioned before, and also on the funeral cloths. The clothes from the tomb (shrouds) and the sudarium are mentioned in the inventories in a list with other relics. None of the six texts that speak about the funeral linens in the Boucoleon palace mention an image of Christ on them[60]. They systematically had no leading role and no image on

---

[55]Robert de Clari. (1924). *La Conquête de Constantinople*. Les Classiques Français du Moyen Age. Paris. pp. 81–82 and 90.

[56]Barta, C., Sabe, P., Orenga, J. M. (2018). *The Beirut icon and the Shroud*. Shroud Newsletter, N. 89, pp. 49–71.

[57]For a thorough study of the Christ relic collection see Bacci, M. (2003). *"Relics of the Pharos Chapel: A View from the Latin West"* in Lidov; Lidov, Alexei, ed., Eastern Christian Relics, Moscow. pp. 235–246. https://www.academia.edu/913214/Relics_of_the_Pharos_Chapel_A_View_from_the_Latin_West. See also Lidov, A. (2012). *"A Byzantine Jerusalem: The Imperial Pharos Chapel as the Holy Sepulchre"* in Jerusalem as Narrative Space. Erzählraum Jerusalem. edited by Annette and Gerhard Wolf Hoffmann. Leiden-Boston. pp. 63–104. Martín J. L., (1987). *Las Cruzadas*. Cuadernos historia 16. n. 140. p. 28.

[58]Durand, J., et al. (2001). *Le trésor de la Sainte Chapelle*. Paris. Louvre. pp. 32–33. William of Tyre. Published by Savio. (1960). Sindon. N. 3. p. 29. Dubarle, A.-M. (1985). *Histoire ancienne du linceul de Turin jusqu'au XIII siècle*, pp. 40–41.

[59]de Clari, Robert. (1924). *La Conquête de Constantinople*. Les Classiques Français du Moyen Age. Paris. p. 82.

[60]Durand, J., et al. (2001). *Le trésor de la Sainte Chapelle*. Paris. 2001. Louvre p. 29, 32, 33, and 35. Accounts of Alexis Comnène, Anonymous Mercati, Reliquiae Constantinopolitanae, Nicholas de Thingeyar, Descriptio sanctuarii and Nicholas Mesarites in Latin, Greek and French dated from the end of 11th century to the beginning of the 13th century.

them is mentioned at all. We might conclude that in this chapel there were cloths from the tomb (parts of a shroud, sudarium and perhaps bandages) without any image. Therefore, this is the first reason to think that the "cloth of the tomb" in Pharos Chapel inside the Boucoleon complex was not the Shroud of Turin. Furthermore, several texts describe these cloths still in the Imperial Complex as "part of" the shroud, not as the whole shroud. This chapel, where the collection was stored, was preserved during the attack in 1204.

Between 1239 and 1242, the Latin emperor Baldwin II sent the following group of 22 relics from Constantinople to his relative, Louis IX of France:

1. The crown of thorns as the most valuable
2. A piece of the cross
3. Blood of Christ
4. The nappies of the infant Jesus
5. Another piece of the cross
6. Blood from a picture of Christ
7. The chain
8. Sacred cloth inserted in a picture (Mandylion)
9. Stone from the tomb
10. Milk of the Virgin Mary
11. The spear
12. A victorious cross
13. The purple mantle
14. The reed
15. The sponge
16. **A part of the shroud (sudarii) in which Christ's body was wrapped in the sepulchre**
17. The towel used to dry the Apostles' feet
18. Moses' rod
19. A part of John the Baptist's head
20. St Blas' head
21. St Clement's head
22. St Simeon's head.

Baldwin took the Byzantine reliquaries that he had in the chapel of Pharos in the Great Imperial Palace of Boucoleon and sent them to Louis, the king of France who commanded to build the *Sainte Chapelle* of Paris to house these relics.

Soon after, Louis sent samples of his relic collection to several famous churches. One of them is Toledo (Spain). Small parcels were sent in 1248 as a gift[61]. They have been preserved in the cathedral until today. This event is well documented. Together with the relics, Louis sent a letter with a list of the collection[62]. It consisted of:

1. A piece of the cross (*de ligno crucis Domini*)
2. A thorn of the crown (*spinis sacrosantae*)
3. Milk of the Virgin Mary (*de lacte gloriosae Virginis Beatae Mariae*)
4. The purple mantle (*de tunica Domini*)
5. The towel used to dry the apostles' feet (*de linteo quo precinxit se Dominus*)
6. A part of the shroud in which His body lay in the tomb (*de sindone, qua corpus ipsius sepultum iacuit in sepulchro*)
7. A part of the Savior's nappies (*de pannis infantiae Salvatoris*)

In the same letter, he specifies the origin of the relics: all of them come from the **imperial treasure of Constantinople** (de *thesauro imperii constantinopolitani*). Therefore, we can accept that part of the relics, which were first in Constantinople and then in the Sainte Chapelle in Paris, are now present in Toledo.

Among the relics there was a part of the shroud (*syndone*) with which Jesus Christ's body was wrapped in the tomb. It is almost the same expression used by Baldwin in the authentication letter but with the substitution of *Sudarii* for *Sindone*. Daniel Duque, César Barta, José Sancho and Felipe Montero (see Fig. 1.25), members of the *Centro Español de Sindonología* (CES), got access and studied all these relics in June 1998[63]. The sample of the

---

[61]Cardenal Lorenzana. (1793). *Patrum toletanorum quotquot exstant opera*. t. III. Madrid.

[62]Solé, M. (1978). *A present of S. Luis, King of France to the Cathedral of Toledo*. II International Congress of Turin. Ed. Paoline, 1979. P. 391.

[63]Duque & Barta. *La Sábana Santa entre Constantinopla y Toledo*. Linteum 26, 27 and 28 (1999–2000).

shroud is a taffeta of 26 threads/cm in weft and 33 threads/cm in warp and S spun, made of linen (Fig. 1.26).

Figure 1.25 Members of the CES in the Ochavo (Cathedral of Toledo).

Figure 1.26 *Sindone* preserved in the Cathedral of Toledo.

It couldn't possibly be a piece of the Italian Shroud, which is a herringbone 3:1, 26×39 threads and Z spun[64]. The fabric found in the reliquary of Toledo, which came from Constantinople, does not belong to the cloth of Turin. Emperor Baldwin had a linen cloth that had been woven in the simplest way in taffeta, incompatible with the one in Turin. The most detailed description of the funeral cloth of the Great Imperial Palace includes the qualifying adjective of "cheap"[65], which fits well with the piece found in Toledo that is a taffeta apparently without any treatment. The Shroud of Turin is herringbone of a much more complex manufacture.

The conclusion is that in the Pharos chapel inside the Great Imperial Palace in Constantinople was the most important collection of Jesus Christ relics. Baldwin sent this collection to Louis IX who housed it in the Sainte Chapelle. The French king sent parts of eight of these relics to Toledo which have been analysed. By this way, currently we know a part of what was there in Constantinople in the Pharos chapel. The funeral cloths of Christ with no image mentioned as *"de Sindone dni"* does not belong to the Shroud of Turin. If the Shroud of Turin was in Constantinople, as certain pollens seem to show, it should rather be identified with the *Sindone* with an image in Saint Mary of Blachernae that was probably taken by the Crusaders and brought in 1205 to Athens[66]. This one must be a different cloth from the funeral cloths that stayed in the Boucoleon palace.

Coming back to the subject of this book, we note that in the collection at Constantinople was included a sudarium[67]. As we will show, at the same time there were already references to a Sudarium in Oviedo. We will deal with this issue in the later chapters.

Some might be astonished to know that there was more than one shroud or more than a single sudarium, but this has

---

[64]Schwalbe, L. A., and Rogers, R. N. (1982). *Physics and Chemistry of the Shroud of Turin*. Analitica Chimica Acta. Vol. 135. p. 41.

[65]Durand, J., et al. (2001). *Le trésor de la Sainte Chapelle*. Paris. Louvre. p. 29.

[66]Cartularium Culisanense translated in Wilson I. (1998). The Blood and the Shroud. London. p. 322.

[67]The Icelandic pilgrim Soemundarson Nicholas list a shroud and a sudarium as two different cloths. Conde Paul Riant: *Exuviae sacrae Constatinopolitanae, II,* pp. 213–216

happened more often than expected. To illustrate this subject, we can now visit the Holy Chamber of the Oviedo Cathedral where the Sudarium is present today. In this chapel there is also a good collection of other assumed relics of Christ and saints. Figure 1.27 shows the old arrangement of the Holy Chamber and to the right, a shelf with some interesting relics (Fig. 1.28). In the lower part, there is a part of the "shroud" (Fig. 1.29). A simple analysis of it shows that it does not belong to the Shroud of Turin.

Figure 1.27 *Camara Santa* (Holy Chamber) of Oviedo.

This introduction shows how complex can the honest study of the relics be. As we said above, there are many relics that claim to be from Jesus Christ. The cross, the Title of the cross, the nails, the thorns of the crown, the blood, the burial cloths, the tunic, the column of scourging, the sepulchre stone, etc. We are almost sure that not all can be relics of the first class, which means, directly belonging to Jesus Christ or used by Him. For example, there are pieces of the cross which come from different species of wood (pine, oak, cedar, etc.). Some of them could be relics of the higher class, which means, they had contact with more direct relics[68]. The most critical authors conclude that all relics are fakes. However, this reasoning is not correct at all. As the late Jesuit, Jorge Loring, said the fact that there are fake dollars does not mean that all the dollars are a fake. Even so an analysis is necessary to determine if a dollar is false or not. The Sudarium of Oviedo allows us to perform this scientific analysis.

Figure 1.28 Right showcase in the *Camara Santa*.

---

[68]First-Class Relics of Christ are directly associated with His life. Second-Class Relics are items owned or frequently used by Him. Third-and higher orders Class Relics are any object put in contact with more direct relics.

Figure 1.29 Part of the shroud preserved in the Camara Santa in Oviedo. (© E. López).

However, if the research is not thoroughly accepted as conclusive, it is more related to the personal convictions than to the scientific evidence. We should not forget that Jesus Christ is the target. For many believers He is alive. For others, He is a fictitious character invented thousands of years ago. For the former, the Shroud and other relics are a source of knowledge of their loved God and they select the supporting evidence disregarding the difficulties. For the second group, the acceptance of the authenticity of some relics can imply changes in their deeper life options leading to daily life changes and it creates an unconscious resistance to free reasoning. Therefore, the subject of this book series is passionate. It is not easy for the two groups to keep within the boundaries of scientific reasoning.

Chapter 2

The History of the Sudarium

Before going into the subject of the history of the Sudarium in this chapter, we should remember that there is a multiplicity of relics and it can be so in the case of the Sudarium. We must clarify that there was a Sudarium in Jesus Christ's tomb and there is a Sudarium in Oviedo Cathedral. In this book there is an assumption that both can be the same. But it is an assumption that needs to be supported by evidence.

The history of the Sudarium begins with the references to the Sudarium of Christ, then continues with references to several "sudariums" and ends with the references to the Sudarium of Oviedo. While discussing some of the references, we will remark on the possible relationship with the Sudarium of Oviedo.

2.1 Tracking of the Sudarium Found in the Sepulchre

The history of the Sudarium starts in the sepulchre of Jesus. As we saw in the previous chapter, the gospel mentions *the Sudarium that had covered his head* or *had been on his head*. Here the key point is the verbal tense. The Sudarium had been on the head of Jesus in a previous moment. The expression allows the

The Sudarium of Oviedo: Signs of Jesus Christ's Death
César Barta
Copyright © 2022 Jenny Stanford Publishing Pte. Ltd.
ISBN 978-981-4968-13-3 (Hardcover), 978-1-003-27752-1 (eBook)
www.jennystanford.com

thought that the Sudarium was on his head during the burial as well as before the burial. The research of the Sudarium of Oviedo leads to the conclusion that the Sudarium was used on the cross and until the moment of putting the corpse in the sepulchre. Later, it was removed from the head, embalmed with unguents, and left near the corpse. This case corresponds to the optional interpretation of the use before the burial. In the opposite interpretation that the Sudarium remained on the head of the corpse during the burial, an additional activity must be hypothesized to correspond to the Gospel; that is, after the deceased changed his previous situation, somebody (maybe himself) removed the Sudarium from the head and put it apart as we established in the previous chapter.

In conclusion, the simplest interpretation of the gospel is compatible with the characteristics of the Sudarium of Oviedo.

Next, we wonder what happened to the Sudarium of the gospel. It was found by the apostles in the sepulchre together with the Shroud. Some texts claim to inform about the location of the burial cloths as either the Shroud or the Sudarium or both.

As Mark Guscin asserts[1], the first reference to the burial cloths after the Gospel of St John can be found in the Gospel of the Hebrews, an apocryphal work that is only known from quotations in other writers—the original was lost. This Gospel existed in the middle of the second century, and it is quite possible that it was written much earlier.

In his *De Viris Illustribus* 2, Jerome quotes the following passage from the Gospel of the Hebrews:

> *Evangelium quoque, quod appellatur secundum Hebraeos, post resurrectionem Salvatoris refert: Dominus autem cum dedisset sindonem servo sacerdotis, ivit ad Iacobum et apparuit ei.*

The gospel called *According to the Hebrews*, recounts this after the resurrection: "But the Lord, after giving the Shroud to the priest's servant, went to James and appeared to him".[2]

We also do not know who the servant of the priest was or who the priest himself was. In any case, in this text only the Shroud is mentioned.

---

[1]Guscin, Mark. (2016). *The Tradition of the Image of Edessa*. Cambridge Scholars. Thesis. p. 194.

[2]ibid.

2.2   Life of Santa Nino of Georgia

After the previous reference that can be dated even before the middle of the second century, we found a reference in the *Life of Santa Nino of Georgia*, a woman who lived before the discovery of the Holy Sepulchre by Constantine, the Roman Emperor. She died in the year 338. Her mother, Susanna, was a sister of the Patriarch of Jerusalem. In the Life of Santa Nino written before the end of the fifth century[3], it is related how the burial cloths were found and kept finally by St. Peter. Guscin gives the translation:[4]

> And they found the linen early in Christ's tomb, whither Pilate and his wife came. When they found it, Pilate's wife asked for the linen, and went away quickly to her home in Pontus; and she became a believer in Christ. Sometime afterwards the linen came into the hands of Luke the Evangelist, who put it in a place known only to himself.
>
> Now, they did not find the **sudarium**[5], but it is said to have been found by Peter, who took it and kept it, but we do not know if it has ever been discovered.

This available version does not clarify if Peter found the cloths after they were hidden by Luke or if Peter found and kept the Sudarium or the Shroud or both. In any case, it says that the place of the Sudarium was not known by the writer at these early centuries.

2.3   Nonnus of Panopolis

Chronologically, the following text already deals specifically with the Sudarium of Christ. It is the paraphrase of the Gospel of St John made by Nonnus of Panopolis in Egypt, written in the first half of the fifth century. He adds a few revealing details to the evangelical account of John 20:6–7.

---

[3]Guscin, Mark. (2016). *The Tradition of the Image of Edessa.* Cambridge Scholars, 2016. Thesis. pp. 194–195.

[4]The life of St. Nino'. *Studia Biblica et Ecclesiastica 5* (1903), p. 11, translation from the Georgian by M. Waldrop & J.O. Waldrop of Oxford.

[5]The translation by Waldrop says "shroud (*sudari*)" without any further explanation.

When Simon Peter arrived after him (John), he immediately went into the tomb. He saw the linens there together on the empty floor; and the cloth that covered his head, with a knot toward the upper back of the part that had covered the hair. In the native language of Syria, it is called sudarium. It was not with the funerary linens, but was rolled up, twisted in a separate place[6],

Other alternative translation is provided by M.A. Prost[7]. We show it together with the original Greek[8].

6 Εσπόμενος δέ πόδεσσιν ὀπίστερος ἵκετο Σίμων,
  καὶ ταχὺς ἔνδον ἵκανεν. ὑπὲρ δαπέδοιο δὲ γυμνοῦ
  σύζυγας ἀλλήλοις λινέους ἐνόησε χιτ ῶ νας,
7 καὶ κεφαλῆς ζωστῆρα παλίλλυτον ἅμματι χαίτης,
  σουδάριον τόπερ εἶπε Σύρων ἐπιδήμιος αὐδὴ,
  οὐ ταφίαις ὀθόναις παρακείμενον, ἀμφιλαφῆ δὲ
  μουναδὸν αὐτοέλικτον, ὁμόπλοκον εἰν ἑνὶ χώρω

6 Then bringing up the rear, old Simon bustled up
  And quickly went inside. There crumpled on the floor,
  They both together recognized the linen cloth
7 And tangled ribbons that had once bound up his hair
  (Which in the tongue of Syria is called Sudarium),
  Not laid beside the winding sheets, but tossed aside
  And wadded up into a bundle by itself.

The additional information that Nonnus provides about the Sudarium is the presence of a knot (ἅμματι). The knot could fix the cloth to the hair or, the knot was in the cloth itself but toward the upper part of the hair. In any case, the sudarium that Nonnus believed to be the Sudarium of Jesus had a knot or had been attached to the hair with a knot. The cloth had been knotted. Maybe Nonnus was witness of the Sudarium he describes or was told about it. The knowledge that Nonnus had, must

---

[6]Bennett, Janice. (2001). *Sacred Blood. Sacred Image. The Sudarium of Oviedo*. Libri de Hispania. Colorado.

[7]Prost, Tony. (1998). *Paraphrase of the Gospel of John* translated into English by Mark A. Prost, Jr. S. Diego, MMII. The Writing Shop Press. Chapter 20. p. 211. http://www.textexcavation.com/documents/nonnosgosjn.pdf.

[8]Nonni Panopolitae. (1834). Metaphrasis Evangelii Ilannei; recensvit. lectionumque varietate instruxit Franciscus Passovius. Accessit Evangelium Sancti Ioannis. Lipsiae.

come from his whereabouts. He knew Beirut and Tyr well and wrote in Alexandria[9], then he had to travel by the Holy Land. It is possible that he visited Jerusalem in Palestine[10]. This supports the probable preservation of the Sudarium in the Near East. He knew how the claimed Sudarium of the gospel was used. This description exactly fits the Sudarium of Oviedo that had also been knotted.

In November 2006, members of the international EDICES team carried out a thorough inspection of the Sudarium in situ. In Fig. 2.1, the author of this book explains the main features of the Sudarium to the Italians members.

Figure 2.1 Inspection 2006. Barta explains the features of the Sudarium ready to be scanned. From left to right: García Iglesias, Barberis, Barta, Montero and Ballosino.

Figure 2.2 shows the wrinkles present in the right-side upper corner with the contrast enhanced. The picture was taken during the inspection of the Sudarium in 2006 when John Jackson's team scanned it with a laser to highlight the wrinkles.

---

[9]Hernández de la Fuente, David. (2008). *Bakkhos anax: un estudio sobre Nono de Panopolis*. Nueva Roma. CSIC. p. 28.

[10]Guscin, Mark. (2006). *La Historia del Sudario de Oviedo*. Ayuntamiento de Oviedo. Avilés. p. 56.

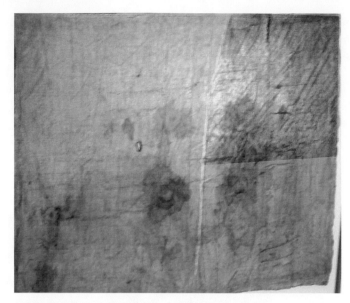

Figure 2.2 Wrinkles of the knot (right side up corner).

The explanation of these wrinkles is the knot made at the edge of the Sudarium of Oviedo in the last phase of its use for fixing the cloth around the head of the corpse (Fig. 2.3). The knot had to remain unchanged for a long time to mark the wrinkles so notably as to be perceived even today. The Sudarium of Oviedo was removed from the head of the corpse maintaining the knot.

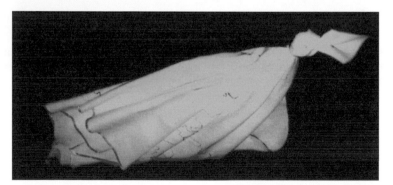

Figure 2.3 Final configuration assumed for the Sudarium of Oviedo with the knot in an edge (© CES).

Thus we can conclude that, in the fifth century the Sudarium, believed to be that of the Gospel and preserved in or near the Holy Land, had a knot just as the Sudarium of Oviedo. This clue leads to the identity of both Sudariums.

2.4   Anonymous Pilgrim from Piacenza

The following chronological references tell us where the Sudarium was. They support the assumption of the previous paragraph that the Sudarium was preserved in the Holy Land. The first witness of the preservation of the Sudarium of Christ is dated about the years 560–570. It is the chronicle made by an anonymous pilgrim from Piacenza, Italy, of the pilgrimage to the Holy Land under the patronage of Saint Antonino of Piacenza. According to the document, in a cave on the banks of the Jordan River was "the Sudarium that was on the head of the Lord".[11]

The introduction of this text in the history of the Sudarium came up in 1998 when the main expert in this subject, Mark Guscin, realized the relationship while exchanging messages with me. I have saved the message in which he thanks me for having found the "missing link". It was a message in Spanish dated 14 October 1998 where he expresses his satisfaction for my finding. I am proud of this contribution to the history of the Sudarium.[12]

The chronicle describes the places visited by the group of pilgrims. It had been written after coming back home. The descriptions provided showed that the travel happened about the year 570[13]. When the pilgrims finished the visit to the Baptism site, to the "other side" of the Jordan River, they entered a cave with seven cells for seven virgins. They prayed but they did not see the face of the virgins. They were told that the Sudarium that had been over the head of Jesus was there. This is

---

[11]Nicolotti, Andrea. (2016). *The Shroud of Oviedo: ancient and modern history.* Territorio, Sociedad y Poder, n° 11, pp. 89–111.

[12]However, the reference had been associated previously by Briansó Augé, Javier. (1997). *El Santo Sudario de Oviedo de la Catedral de Oviedo.* Oviedo. p. 60.

[13]Guscin, Mark. (2006). *La Historia del Sudario de Oviedo.* Ayuntamiento de Oviedo. Avilés. pp. 56–57.

the same expression used in the gospel of John. Guscin gives the Latin original text:[14]

> *In ipsa vero ripa Jordanis est spelunca, in qua sunt septem cellulae cum septem puellis, quae ibi infantulae mittuntur: et cum aliqua ex eis mortua fuerit, in ipsa cellula sepelitur; et alia cellula inciditur, et alia ibi mittitur puella ut numerus stet, et habent foris qui eis cibaria praeparet. In quo loco cum magno timore ingressi sumus ad orationem, nullius ibi faciem vidimus. In ipso loco dicitur esse <u>sudarium quod fuit super caput Jesu</u>. Non multum longe a Jordane, ubi baptizatus est Dominus, est monasterium sancti Johannis valde magnum.*

Even if in some cases the word sudarium is used to describe the shroud, in this case confusion is discarded, according to Guscin. When a text makes the mistake of naming a "sudarium" a shroud, it says that the cloth was over the head of Jesus "in the sepulchre"[15]. The lack of the typical words "in the sepulchre" in Saint Antonino's text proves that the author did not speak about the shroud.

There is also another version of the Latin text.

> *In illa ripa Iordanis est spelunca, in qua sunt cellulae cum septem virgines, quae ibi infantulae mittuntur, et dum aliqua ex ipsis mortua fuerit, in ipsa cellula sepelitur et alia cellula inciditur et mittitur illic alia infantula, ut numerus stet, et habent foris, qui eis permanent. In quo loco cum timore magno ingressi sumus ad orationem, faciem quidem nullius videntes. In ipso loco dicitur esse sudarium, qui fuit in fronte Domini"*[16].

We can analyse the differences. The first difference is that the seven persons that live in the cave are called *puellis* or *virgins*. There is a whole consensus in considering them maidens or nuns, females in any case. The other interesting difference is about the Sudarium covering the <u>head</u> or the <u>front</u> (or face). We have already discussed this topic in Chapter one (Section 1.2.2, On the face). The use of the Sudarium covering the face that the

---

[14]Guscin, Mark. (2006). *La Historia del Sudario de Oviedo*. Ayuntamiento de Oviedo. Avilés. Appendix I. p. 195.

[15]Guscin, Mark. (2006). *La Historia del Sudario de Oviedo*. Ayuntamiento de Oviedo. Avilés. p. 57.

[16]Text ed. Milani, 1977 provided by Nicolotti, Andrea. (2016). *The Shroud of Oviedo: ancient and modern history*. Territorio, Sociedad y Poder, n° 11, p. 91.

writer of the anonymous of Piacenza could understand, excludes the identification of the Sudarium with a chin band or a headdress.

A translation to English is provided by A. Stewart[17]

> On the very bank of Jordan is a cave, in which there are seven cells with seven maidens, who are sent thither when young children; and when any one of them dies, she is buried in her own cell, and another cell is cut out and another child is placed in it, that the number may be maintained. They have persons without the cave, who provide them with food. Into this place we entered with great awe to pray. We saw no one's face there. In this place there is said to be <u>the napkin that was about the head of Jesus</u>. Above the Jordan, and not far from the river where our Lord was baptized, is the Monastery of St. John.

Hermit cells overlooking the river Jordan

Figure 2.4 Hermit cells. (Photo courtesy Rustom Mkhjian of the Baptism Site commission).

The archaeological excavations at the place of the Baptism of Jesus started in 1997 and it was opened for tourism only

---

[17]Stewart, Aubrey. (1887). *Of the Holy Places Visited by Antoninus Martyr*. Palestine Pilgrims Text Society. London. p. 11. An alternative translation is provided by Guscin, Mark. (1999). Recent Historical Investigations on the Sudarium of Oviedo. Proceedings of the 1999 Shroud of Turin International Research Conference. Richmond. Virginia. (Antoninus Martyr).

some years ago. The site has been nominated to be Human Patrimony. In August 2013 a team of the Centro Español de Sindonología (CES) composed of the President, Jorge Manuel Rodriguez Almenar, the Director of the Research Teams, Alfonso Sánchez Hermosilla and me, visited the archaeological site where some caves remain today. Our host was Eng. Rustom Mkhjian, Assistant to the General Director of the archaeological Baptism Site. He was kind enough to lead us to one of the caves (Fig. 2.4 and Fig. 2.5).

Figure 2.5 César Barta (left), Rustom Mkhjian (right) and Jorge M. Rodríguez (middle) in a cave near the River Jordan.

The description of Antonino's Pilgrim does not match with the caves we saw. The caves for the seven virgins should be all together and around a common place. In the Evaluations of Nominations for the World Heritage in 2015, the ICOMOS report includes the assessment of the Baptism Site "Bethany Beyond the Jordan"[18]. On page 50 the report says:

> A cluster of monk caves carved into the Qattara Hills, also called hermit cells, is located at 300 metres distance to the Jordan River. These caves were in the past accessible from the western and

---

[18]International Council on Monuments and Sites: Home (ICOMOS) 2015 Evaluations of Nominations of Cultural and Mixed Properties. Report for the World Heritage Committee. pp. 49–60. http://documentation.icomos.org/whu/Evaluations-2006-/EN%20Volume%20d'%C3%A9valuation%202015%20COMPLET.pdf

south-western sides by ropes, ladders or staircases which no longer exist. Semi-circular niches are carved into the eastern wall of each cave, which was divided into two rooms, assumed to have been reserved for praying and for living.

A number of tombs were identified adjacent to or within the churches. They seem to be burial places of monks or individuals closely associated with the churches. Most graves are rectangular and adequate in size to receive a single outstretched body. The burials have been dated to the Byzantine and early Islamic periods (5th–7th century CE). The property revealed archaeological finds, including coins and ceramics and serves as an epigraphic reference.

To access the available cave now, wooden stairs have been installed. Although archaeologists have found tombs, they have not yet been able to identify it as the actual place of the caves for the seven virgins. It remains a challenge for future archaeological research. However, each individual cave could be like the cave we visited. We took some samples for further analysis.

In conclusion, before the invasion of Chosroes II (year 614) the assumed Sudarium of Christ was at the East of the Jordan River worshiped by Christians. The fact that the pilgrims did not see it, support the assumption that it was saved in a chest or reliquary.

2.5   Mentions of Other Sudariums

There are several brief mentions of sudarium, barely descriptive, and also those that name a cloth as sudarium by mistake.

An example of the last case is the episode mentioned in the book entitled *Arculfi relatio de locis sanctis, ab Adamanno scripta* that involves the Gallic bishop Arculf. He traveled from France to the Holy Land. After he returned from his pilgrimage, he visited the Scottish island, Iona where he was hosted by the abbot Adamnan. Listening to Arculf's stories, Adamnan wrote a narrative of his pilgrimage to the Holy Land. The travel is dated around the year 679[19]. The account tells that Arculf was in Jerusalem

---

[19]Bennett, Janice. (2001). *Sacred Blood. Sacred Image. The Sudarium of Oviedo*. Libri de Hispania. Colorado. p. 26. Also, Guscin, Mark. (2006). *La Historia del Sudario de Oviedo*. Ayuntamiento de Oviedo. Avilés. pp. 62–64.

and describes the trial of the cloth between Jews and Christians through a fire test. The Sudarium was taken out of a sanctuary and shown to the crowd; he had even been permitted to kiss it. But the conclusive data is the size he gives of the cloth. He reported that the sudarium was 8 feet long (2.4 meters). This is a proof of the confusion of terminology because a cloth of such dimension is clearly a shroud. A sheet so big cannot be used for wrapping only the head, but it would be used to cover the whole body. However, the expression in the text of Arculf that is available to us uses "sudarium":

> *De illo quoque sacrosanto sudario quod in sepulchro Domini super caput ipsius fuerat positum. [...] in scrinio ecclesie in alio involutum linteamine recondunt.*
>
> *Quod noster frater Arculfus alio die de scrinio [e]levatum vidit et inter populi multitudinem illud osculantis, ipse osculatus est in ecclesiae conventu, mensuram longitudinis quasi octonos habens pedes[20]*

Thanks to this text, we must keep in mind that, when we read "sudarium" in a Latin old text, it can refer either to the Shroud or to the Sudarium. Moreover, a detail that we should not pass over is the fact that the relic was kept in a box (reliquary or shrine) wrapped in another linen. This should be the way Christians safeguarded their relics in the antiquity.

According to several authors, the cloth kissed by Arculf ended in France. It could be the shroud of Cadouin[21] which is 2.80 metres long. This was examined by experts who concluded that it cannot be a shroud related to Christ because its inscription is actually worshipping Allah. The other candidate was the shroud of Compiègne[22] which was 2.40 metres long. It existed until the French Revolution when it was destroyed.

We could conclude that the only direct witness of a place where the Sudarium of Christ was preserved is that of the banks

---

[20]Arculfi relatio de locis sanctis scripta ab Adamnano. c670. I. Pomialovky. Book 1, XI. pp. 11–14.

[21]Natale Noguier de Malijay. (1986). *The Cloth of Cadouin*. Shroud Spectrum International No. 18 Part 5. pp. 12–17.

[22]Alarcón Benito, J. (1994). *La Sábana Santa. El Gran Misterio del Cristianismo*. Temas de Hoy. pp. 234–236. Also, Kersten and Gruber referred by Bennett, Janice. (2001). *Sacred Blood. Sacred Image. The Sudarium of Oviedo*. Libri de Hispania. Colorado. p. 26.

of the Jordan river. And it happened before the fall of Jerusalem under the attack of Chosroes in the year 614. As centuries passed, the mention of relics increased. As already stated, when the original word is sudarium, it could not be just related to the napkin of the gospel or to the Sudarium of Oviedo. It could be indicating a shroud. It happens particularly when the text is written in French where the shroud is named *suaire*. According to Guscin[23], the word 'sudarium' was often used to mean a shroud, the large cloth that covered the whole body. The expression "the sudarium that was on his head in the sepulchre" in those centuries would normally be a reference to the Shroud. Hence, we will only list below the references where sudarium is mentioned together with the shroud. In such cases, it is clear that the sudarium is different from the shroud. Moreover, we must bear in mind that many mentions of these relics only refer to a small part of the supposed shroud or sudarium. We have seen in the previous chapter that there is part of a shroud in Oviedo cathedral and some bits of relics in Toledo cathedral.

Hence, many references of sudarium without much description have not been considered here below[24]. For example, the supposed relic of the sudarium in the Sancta Sanctorum of the church of St John Lateran[25] could be only a cloth fragment among the 88 small relics that were in the Pope's chapel. Only four were textile

---

[23]Guscin, Mark. 1998. The Oviedo Cloth. The Lutterworth Press, p. 11 and personal communication. E-mail from Guscin to Barta, 24 October 1998. Subject referencias sudario. Also, Guscin, Mark. 2006. *La Historia del Sudario de Oviedo*. Ayuntamiento de Oviedo. Avilés. p. 57.

[24]For example, the reference of the year 808 to the sudarium in the Holy sepulchre provided by Andrea Nicolotti. 2016. *The Shroud of Oviedo: ancient and modern history*. Territorio, Sociedad y Poder, n° 11, pp. 91–92. It only says *In sancto sepulchro Domini [...] sanctam crucem et sudarium*. Tobler, Titus. (1874). *Descriptiones Terrae Sanctae ex saeculo VIII. IX. XII. et XV*. Leipzig 1874. J. C. Hinrichs 'sche Buchhandlung. Most of the references given by Nicolotti deal only with a cloth and can be just the shroud and not the sudarium.

[25]Nicolotti, Andrea. (2016). *The Shroud of Oviedo: ancient and modern history*. Territorio, Sociedad y Poder, n° 11, pp. 93–94. "sudarium quod fuit super caput eius". The sentence in the list of John the Deacon is larger: Sudarium quod fuit super caput ejus, quod est unum de quinque linteaminibus, quibus sanctissimum corpus cjusdcm Domini fuit involulum. (Cloth that was on his head, which is one of five linens, that enveloped his most holy body). Lauer, Philippe. (1906). *Le Trésor du Sancta Sanctorum*. In: Monuments et mémoires de la Fondation Eugène Piot, tome 15, fascicule 1–2. p. 31.

and one of them was of Christ[26]. There is no reliable evidence of the existence of a whole sudarium in the Lateran Palace in Rome. It is not mentioned among the main relics of Christ in the Sancta Sanctorum[27].

On the other hand, there are relics whose achievement is attributed to Charlemagne. One of them is the shroud of Compiègne, already mentioned above. It was originally part of the treasure that Charlemagne offered to his Palatine Chapel in Aachen. Today there are four relics venerated in Aachen and none is either the shroud or the sudarium (St. Mary's robe, swaddling cloths, St John the Baptist's beheading cloth and Jesus' loin (hip) cloth[28]). The sources of Compiègne say that a half of the shroud was offered to Charles, the Bald for the foundation of his monastery in Compiègne, around 875. But it disappeared during the French Revolution. However, Kornelimünster claims to have three cloth relics that came originally from the treasures of Charlemagne, via his son and successor, Ludwig, the Pious, who granted them to Kornelimünster. Today, there are in this city three relics associated with Christ: the towel that Jesus wore to wash and dry the feet of his disciples, the Shroud in which His body was wrapped and the Facecloth that was wrapped around His head.

This is an example where there are, in the same place (Kornelimünster), two cloths purported to be the shroud and the sudarium. However, when we get enough details, we realize that none of them is a sudarium. The so-called shroud in which His body was wrapped is an artistically woven linen cloth, comparable to a decorative blanket (see Fig. 2.6).

It has a width of about 1.80 m and a length of 1.05 m. And the so-called sudarium in Kornelimünster is still bigger! It has a size of 4 m × 6 m. It is a byssus fabric, produced from very fine silk fibers. It is folded 16 times on other red silk support to be saved under a gauze screen, placed for protection (Fig. 2.7).

---

[26]Luchterhandt, Manfred. (2017). The popes and the loca sancta of Jerusalem. Relic practice and relic diplomacy in the Eastern Mediterranean after the Muslim conquest. In: Natural Materials of the Holy Land and the visual translation of place 500–1500, ed. R. Bartal, N. Bodner and B. Kuehnel, London-New York. pp. 36–63. Table 3.2.

[27]Marani, Tommaso. (2012). The Relics of the Lateran According to 'Leidarvisir', the 'Descriptio Lateranensis ecclesiae', and the Inscription Outside the 'Sancta Sanctorun. Medium Ævum. vol. 81. no. 2. pp. 271–288.

[28]https://heiligtumsfahrt-aachen.de/en/the-aachen-pilgrimage/the-relics.

Even if it is named face cloth, its huge size of 6 meters and the nature of the fabric (silk), excludes the possibility that it corresponds to a sudarium. In Chapter one, we have shown why the sudarium of the gospel must have a reduced size, like a towel. The web site of the Catholic Church in Germany makes no claim on the authenticity of any of these relics[29].

Figure 2.6 Shroud of Kornelimünster (© heiligtumsfahrt-aachen.de).

Figure 2.7 Sudarium of Kornelimünster (© heiligtumsfahrt-aachen.de).

[29]https://www.vennbahn.eu/wp-content/uploads/2019/09/Vennbahn-Stories_2_ Kornelim%C3%BCnster_EN.pdf.

These facts lead us to be skeptical when we are informed about a relic of Christ. Not every claimed relic of Christ is authentic, but at the same time it does not mean that every claimed relic of Christ is a fake. Research is necessary to assess and reach a verdict.

Figure 2.8 Sudarium of Mainz inside its reliquary (©Winfried Wilhelmy. In Gold geschrieben).

Other hardly-known purported sudarium of Christ is preserved today in the crypt of the cathedral of Mainz[30] (Fig. 2.8). The origin of this sudarium is attributed to the life of St. Bilhildis who received it from a French princess[31] at the end of 7th century

---

[30]Brodehl, Thomas. (2009). *Die heilige Bilhildis und das Schweißtuch Christi zu Mainz*. In: Johannes A. Wolf (Hrsg.): Der schmale Pfad. Orthodoxe Quellen und Zeugnisse. Band 29. S. pp. 94–115.

[31]Imnechild or Kunegundis, more data not available.

or beginning of 8th century. According to Matheus[32], this cloth is part of the "sudarium" of Kornelimünster that we saw above being more than 6 meters long. However, although it is also described as a byssus fabric, it seems to be an apparent contradiction because in this case of Mainz the textile is linen and in the case of Kornelimünster it seems to be silk. But, according to Wilhelmy[33], there is no contradiction because the structure of this linen fabric and its sheen could lead to assume that it was silk. In any case, it is almost transparent with few threads per square centimeter (Fig. 2.9). This sudarium does not have bloodstains and the possible scientific analysis seems limited.

Figure 2.9 Sudarium of Mainz (©Winfried Wilhelmy. In Gold geschrieben).

The documents that explain the origin of the relics of Kornelimünster and Mainz are from centuries after the event.[34]

---

[32]Matheus, Michael. (1999). *Pilger und Wallfahrtsstätten in Mittelalter und Neuzeit.* Franz Steiner Verlag Stuttgart. p. 107.

[33]Winfried Wilhelmy und Tino Licht (Hg.). (2017). *In Gold geschrieben. Zeugnisse frühmittelalterlicher Schriftkultur in Mainz.* Bischöflichen Dom- und Diözesanmuseums. Mainz; Bd. 9. Regensburg. p. 35.

[34]Nicolotti, Andrea. (2016). *The Shroud of Oviedo: ancient and modern history.* Territorio. Sociedad y Poder. n° 11, p. 92. Note 10.

To classify the sources by its reliability, we can differentiate between the documents that describe an event contemporary to the original writing (more reliable) and the documents that talk about event in the past transmitted through generations, usually introducing fantasies and myths.

With this premise in mind, we look at other references to a sudarium in the past.

Ancient texts mention the existence of cloths claimed to be the Sudarium of Christ but those have not been related to the Sudarium of Oviedo. For example, we learned that there was a sudarium in Constantinople when the Holy Ark was already in Oviedo with perhaps the Sudarium inside. Michele Bacci[35] gives the list of pilgrims that saw a sudarium in Constantinople. Only two indicated the two cloths (shroud and sudarium) as different relics[36]. At the end of the 11th century, sudarium and shroud are mentioned in Constantinople in the Pharos chapel: *lintheamen et sudarium sepulture eius*[37], (linen cloth and sudarium of his burial). But the most decisive is the reference of Nicholas Soemundarson who wrote, in his native Icelandic, an inventory of the Byzantine city relics seen during his 1157 visit. Often, the previous references to Soemundarson use the Latin version provided by Count P. Riant[38]. However, thanks to the Icelandic teacher, Ásta Svavarsdóttir, we have a reliable translation from the original language. In the ancient palace there were *líkblæjur med sveitadúkur ok blóði Krists*. The analysis of Svavarsdóttir is that *líkblæjur* is the feminine plural of "líkblæja", meaning 'a winding-sheet', i.e., a (thin) cloth used to wrap around a corpse. *Sveitadúkur* is masculine

---

[35]Bacci, Michele. (2003). Relics of the Pharos Chapel. A View from the Latin West, in: Eastern Christian Relics. Ed. Alexi Lidov. Moscow. pp. 243–244.

[36]Till about 1150 the sudarium is described using the same expression as in the gospel: Paul Riant, Exuviae sacrae Constantinopolitanae, 2 vols (Geneva 1878, 1879). repr. With introduction by Jannic Durand, Paris (2004), vol. 2. p. 211. Durand, and M. P. Laffitte. (2001) *"Le Trésor de la Sainte-Chapelle"*, Publication du Louvre 71 (Paris 2001). p. 52. *Sudarium quod fuit super caput eius* (The sudarium that was upon his head). As we said, in that century, this expression usually refers to the shroud. William of Tyre (1171) only mentioned the *sindonem*. Historia Rerum in Partibus Transmarinis Gestarum. Liber Vigesimus. Chapter 25.

[37]Anonymous (Mercati). (1976), Ed. Ciggar K.N. *Une Description de Constantinople traduite par un pèlerin anglais. REB.* vol. 34. pp. 211–267. J. Durand, and M. P. Laffitte. (2001). *"Le Trésor de la Sainte-Chapelle"*, Publication du Louvre 71. Paris. p. 52.

[38]Paul Riant. (2004). Exuviae sacrae Constantinopolitanae. 2 vols (Geneva 1878, 1879) repr. With introduction by Jannic Durand. vol. 2. pp. 211–212.

singular meaning 'a napkin' (literally 'sweat cloth'), often used to cover the face of a dead person. The last part, "*blóði Krists*", means 'the blood of Christ'. These last words imply that they should be the Passion cloths of the Gospel when the blood of Christ was shed. The original Icelandic word means more properly sheet than bands (*fasciae* used in the translation of the Latin). It is possible that it may denote a large body cloth. The two cloths, thus, correspond to the burial linens. Although, it is possible that the blood was in a separated ampoule, we can safely assume that one or both the cloths were stained with blood.

While the sheet (shroud) can be considered the *part* of the shroud that arrived in Paris and then a piece of it in Toledo (see Chapter 1), the track of the sudarium of Constantinople is lost. In the list of 22 relics sent by Baldwin II from Constantinople to his relative Louis IX in Paris, there is a *part of the shroud in which Christ's body was wrapped in the sepulchre.* We note that it is a part, not the whole shroud. As we have already seen, a piece of this shroud was sent to Toledo and preserved there until today. Then, we know now how the part of the shroud venerated in Constantinople was. The sudarium of Constantinople was not among the relics sent by Baldwin. The pilgrims visiting Constantinople after the Fourth Crusade do not mention a sudarium among the relics remaining in the city. If there was a sudarium in the Byzantine capital, we do not know what happened to it. Its track is lost. But it cannot be the Sudarium of Oviedo if this was in the Holy ark in Oviedo when Soemundarson visited Constantinople the year 1157.

2.6 References of the Sudarium of Oviedo

After categorizing the most valid references to sudariums from the first to the twelfth century, we consider the site of the seven virgins near the Jordan River about the year 570 as the most reliable. It does not give an accurate description of the cloth but at least, it is compatible with the true concept of the Sudarium that was on the head of Christ. In addition, we learned that around the Holy Land, the writer Nonnus of Panopolis thought that the Sudarium had a knot.

In the year 1075, on 14 March, a "minutes of meeting" was written certifying what happened the day before, that is on 13 March[39], on the fourth Friday of Lent. This kind of document is scarce in the subject of relics. But in the case of the Sudarium of Oviedo, we have a reliable reference as close as just one day after the event.

The document[40] describes the solemn opening in Oviedo of a Holy Ark in the presence of King Alfonso VI of Leon, his sister Lady Urraca, the *Infanta* Lady Elvira, his knight Rodrigo Díaz de Vivar, El Cid Campeador, several bishops and the high ecclesiastical dignitaries. The presence of those high ranked people is evidence of the great importance that the opening of the chest meant. It is also proved by the previous requirements demanded for opening the coffer. These included the celebration of services, extraordinary fasts, and songs, with censers that impregnated every corner with the fragrance or incense.[41]

The document describes what Holy Ark contained: relics of Christ, relics of his mother as well as innumerable remains of the Holy Saints, Prophets, Martyrs, Confessors, and Virgins. The number of relics was more than eighty in a whole. Seven of the Lord (2 included after those of the Virgin); two of the Virgin; eight of the Apostles; one of the prophets; one of Saint John the Baptist; four of people from the Old Testament and sixty-two of other saints, apart from the final expression of the many others, impossible to count.

The document is a donation of which a copy of the 13th century is found today in folios 1 and 3 of the Gothic Book in the Archive of the Oviedo Cathedral (Fig. 2.10).

Inside the Holy Ark was a Sudarium or a part of it. It can be the beginning of the history of the Sudarium of Oviedo (year 1075). We can follow it from this moment onwards. And from this moment backwards, we can follow the legend of the Holy Ark. This must be clear. There are several documents that refer to the Holy Ark but not directly to the Sudarium of Oviedo. The fact that

---

[39]*III idus martii.* In the document. Third day of *idus* March that means 13th March.

[40]The original Latin transcription and the Spanish translation are in López Fernández, Enrique. 2004. *Las Reliquias de San Salvador de Oviedo*. MADU. Granda-Siero. pp. 223–227.

[41]Bennett, Janice. 2001. *Sacred Blood. Sacred Image. The Sudarium of Oviedo*. Libri de Hispania. Colorado. p. 41.

the Sudarium could be inside the Ark allows us to assume that the history of the Sudarium should be linked to the history of the Ark. However, we do not have evidence of when the Sudarium of Oviedo was introduced in the Ark. Neither can we exclude, as we will see, the hypothesis that the Sudarium was never inside the Holy Ark.

Figure 2.10 Langreo donation parchment (© E. López).

We know some details about how the relics inside the chest were placed. They were found inside their own golden and silver boxes, each one with its label in Latin.[42] They would have been small sized as it is typical for the relic collections in most of the cathedrals and churches. We can assume this small size

[42]López Fernández, Enrique. (2004). *El Santo Sudario de Oviedo*. MADU. Granda-Siero. p. 44. Footnote 50.

because, according to the Cambrai manuscript[43], there were 12 box-reliquaries inside the Ark and there were 13 relics inside one of them.[44] Even though this information is provided years after the opening event, in the year 1588, the data is reliable because it is in the report of the pastoral visit of the bishop Diego Aponte de Quiñones. There were still 12 box-reliquaries inside the Ark. The reliquaries do not have to be those arriving in the Ark, but it gives us an idea of how the relics were stored. Today there are old reliquary boxes preserved still in the museum of the cathedral of Oviedo. Of course, the Sudarium did not come inside the Ark unfolded in its whole dimension. It has several creases which prove that it was folded at least in eight layers presenting a size of 21 cm × 27 cm.[45] This size is even too big to be in one of the reliquary-boxes found inside the Ark. The biggest length once folded (27 cm) is even bigger than the other relic of the shroud that also came inside the Ark, whose unfolded size is smaller than 25×25 cm[46]. This shroud, before being exposed alone in its silver frame as we can see it today, was saved inside one of the box reliquaries together with other 13 relics of Christ[47]. It had to be much folded. The Sudarium shows creases that suggest having been folded in 16 or even 32 layers resulting in a small size of 21×14 cm or 10×14 cm, respectively.[48]

However, the mention of the Sudarium inside the Holy Ark in the referred document of donation of Langreo, only says "*de sudario eius*" that means "a part" of the Sudarium. Moreover, in the donation document there is only one instance of a funeral cloth among more than eighty relics. But in Oviedo there is also

---

[43]The Cambrai 804 manuscript belongs to a group of three manuscripts with the same account. The other two are de Valenciennes 30 and the Brussels II 2544. Guscin, Mark. (2006). *La Historia del Sudario de Oviedo*. Ayuntamiento de Oviedo. Avilés. p. 121.

[44]Guscin, Mark. (2006). *La Historia del Sudario de Oviedo*. Ayuntamiento de Oviedo. Avilés. p. 131.

[45]Montero, Felipe. (2012). *Description of the Sudarium of Oviedo*. Proceeding of the first International Congress for the Shroud. Valencia 28–30 de April 2012. p. 24.

[46]López Fernández, Enrique. (2004). *Las Reliquias de San Salvador de Oviedo*. MADU. Granda-Siero. p. 166. Note 316.

[47]López Fernández, Enrique. (2004). *Las Reliquias de San Salvador de Oviedo*. MADU. Granda-Siero. p. 156.

[48]Ramos, Teresa. (1994). *El Sudario de Oviedo. Otros Elementos*. Proceeding of the 1st International Congress about the Sudarium of Oviedo. Universidad de Oviedo. Oviedo. p. 178.

a part of the shroud. "*De sudario eius*" can refer actually to a part of the *shroud*. If we consider that the mention of *de sudario eius* refers to the Sudarium, then, the list fails to include the part of the *shroud* and vice versa. If we trust the Langreo document, we have to choose between Sudarium and Shroud. We took the criteria to dismiss the references of sudarium when there is only the mention without more description and when the shroud is not mentioned as other coexisting funeral cloth. What is more, as we saw in the previous chapter, there is a part of the shroud preserved until today in the Oviedo cathedral (Fig. 2.11).

Figure 2.11 Part of the shroud preserved in the Camara Santa in Oviedo. (© E. López).

Then the expression "*de sudario eius*" could be assumed to indicate this part of the shroud. Doubt increases when we learn about some documents of the following years that give the list of the relics including the *shroud* (*sindone*) and excluding the *Sudarium*[49]. We have however, other almost contemporaneous document that mentions both shroud and sudarium in the list of relics included in the Holy Ark. It is the Book of Testaments

[49](see later) Corpus Pelagianus, bulletins and documents of Holy Chamber brotherhood only mention "de sindone". Guscin, Mark. *La Historia del Sudario de Oviedo*. Ayuntamiento de Oviedo. Avilés. p. 107, 206. López Fernández, Enrique. (2018). *El Santo Sudario de Oviedo*. Soluciones Gráficas. p. 260. López Fernández, Enrique. (2008). *Historia de un Silencio. Sudario de Oviedo a través de los siglos.* MADU. p. 344.

(where "testament" means donation). This is a collection of documents written by the bishop Pelayo recording all the donations made to the cathedral church of San Salvador in Oviedo[50]. This bishop was in charge of the dioceses of Oviedo between around 1101 and 1129 and two more years in the 1140s. Pelayo mentions, among the relics, "a part of the Lord's shroud" and "a part of the sudarium of the Lord"[51]. We should observe that the shroud in the list of Pelayo is among the first group of relics of Christ corresponding to the sudarium in the list of the donation of Langreo. In this last document (Langreo) the sudarium is named in the 9th place after the tunic divided by lots, and the shroud is not mentioned in any place of the list. In Pelayo's document the shroud is named in the fifth place before the tunic and the sudarium is mentioned in the place 27 out of 28 after the cape of Elijah and before a roasted fish. That means, with no relevance. In addition, it is again "part of" sudarium (*De Sudario Domini*[52]).

This is the list of relics in the donation of Langreo (year 1075)[53]

1. Lord's wood
2. Blood of the Lord
3. Bread (last supper)
4. Sepulchre of the Lord
5. Soil stepped by the Lord
6. Virgin dress
7. Virgin milk
8. Robe of the Lord
9. **Sudarium of the Lord**
10. Relic of Saint Peter
11. Relic of St Thomas
12. Relic of St Bartholomew
13. Relic of St Justus and St Pastor
    etc.

---

[50]Guscin, Mark. (1999). *Recent Historical Investigations on the Sudarium of Oviedo.* Proceedings of the 1999 Shroud of Turin International Research Conference, Richmond, Virginia.

[51]"de sindone Domini" and "de sudario Domini".

[52]Guscin, Mark. (2006). *La Historia del Sudario de Oviedo.* Ayuntamiento de Oviedo. Avilés. p. 201.

[53]López Fernández, Enrique. (2004). *Las Reliquias de San Salvador de Oviedo.* MADU. Granda-Siero. pp. 223–227.

This is the list of relics in the Book of Testaments (around year 1120)[54]

1. Blood of the Lord
2. Lord's wood
3. Sepulchre of the Lord
4. Crown of thorns
5. **Shroud of the Lord**
6. Robe of the Lord
7. Diapers of the manger
8. Breads (multiplication)
9. Bread (last supper)
10. Manna
11. Land of the Mt Olivet
12. Soil from Lazarus resurrection
13. Stone of Lazarus' own grave
14. Virgin milk
15. Virgin dress
16. Chasuble of Ildefonso
17. Hands of Saint Stephen
18. St Peter sandal
19. Forehead of John the Baptist
20. Hair of the Innocents
21. Bones of Azarías, Ananías and Misael
22. Marta and María hair
23. Lord's Tombstone
24. Olives Mt. Olivet
25. Sinai stone
26. Part of the Elijah's mantle
27. **Part of the Sudarium of the Lord**
28. Roasted fish and honeycomb

The *Crónica de Alfonso III* in the *Corpus Pelagianus*[55] follows the Book of Testaments. Only adds the hydra of Cana and three

---

[54]Guscin, Mark. (2006). *La Historia del Sudario de Oviedo*. Ayuntamiento de Oviedo. Avilés. p. 201.

[55]Manuscript 1513. Biblioteca Nacional de Madrid. Guscin, Mark. 2006. *La Historia del Sudario de Oviedo*. Ayuntamiento de Oviedo. Avilés. p. 107, 206.

new relics at the end of the list. However, although it also mentions the part of the shroud in the fifth place, it does not mention the Sudarium at all.

This is the list of relics in the Corpus Pelagianus (around 1120–1130)[56]

1. Blood of the Lord
2. Lord's wood
3. Tombstone of the Lord
4. Crown of thorns
5. **Part of the Shroud of the Lord**
6. Robe of the Lord
7. Diapers of the manger
8. Breads (multiplication)
9. Bread (last supper)
10. Manna
11. Land of the Mt Olivet
12. Soil from Lazarus resurrection
13. Stone of Lazarus' own grave [Hydra of Cana]
14. Virgin milk
15. Virgin dress
16. Chasuble of Ildefonso
17. Hands of Saint Stephen
18. St Peter sandal
19. Forehead of John the Baptist
20. Hair of the Innocents (and fingers)
21. Bones of Azarías, Ananías, and Misael
22. María's hair
23. Sinai stone
24. Part of the Elijah's mantle
25. Roasted fish and honeycomb
26. Chain of S. Peter
27. Crox
28. Hair of John the Baptist

---

[56]Alonso Álvarez, R. (2018). De Cruore Domini: La Reliquia de la Santa Sangre en la Catedral de Oviedo y el Milagro del Cristo de Beirut. Medieval Studies in Honour of Peter Linehan. Edizioni del Galluzzo, p. 54.

There is another document with the list of relics. It is the cover of the Holy Ark. They are letters engraved in metal (difficult to read nowadays). But there are old transcriptions that allow us to have the list. It is similar to the list of the donation of Langreo because it names the Sudarium and does not name the shroud. The Sudarium stands in the fifth place like in the list of the Book of Testaments. *De sepulchro dominico, ejus atque sudario et cruore scisimo.*

This is the list of relics on the cover of Ark (year 1090–1100)[57]

1. Lord's wood
2. Robe of the Lord
3. Bread (last supper)
4. Sepulchre of the Lord
5. **Sudarium of the Lord**
6. Blood of the Lord
7. Soil stepped by the Lord
8. Virgin dress
9. Virgin milk
10. Relic of St Peter
11. Relic of St Thomas
12. Relic of St Bartholomew, etc.

Ambrosio de Morales[58] also gives a transcription of the cover of the Holy Ark. He reads "shroud" instead of "sepulchre": *De **sindone** dominico, ejus atque **sudario** et cruore scisimo*[59]. In this case, we would have reference for the two burial cloths together. However, he is the only one who reads this way, and it is clearly a mistake because, if we read *sindone*, we lack the *sepulchro* which is there in every other document that lists the relics.

It is worth mentioning here the confusion between sudarium and shroud in the case of the report of Ambrosio de Morales.

---

[57]García de Castro, César. (2017). *El Arca Santa de la catedral de Oviedo*. Ars Mediaevalis. p. 84. The year 1113 written in the inscription corresponds to the Medieval Hispanic calendar, which adds 38 years to the current calendar.

[58]Florez, H. (1765). *Viage de Ambrosio de Morales*. Madrid. Antonio Marín. p. 80.

[59]López Fernández, Enrique. (2004). *Las Reliquias de San Salvador de Oviedo*. MADU. Granda-Siero. p. 274. Note 49 and 74. Gómez Moreno and Diego Santos reconstruct the text from the previous work of Ciriaco Miguel Vidal and from references of the 16th century. They read sepulchre and sudarium.

As we will see, this ecclesiastic visited Oviedo in 1572 and he wrote that the inscription around the Holy Ark *said that there is **part of the shroud**[60] of our Redeemer, and there is not the entire **Sudarium**[61] of our Redeemer in this Relic, but part of it, as it was not possible to wrap the entire Divine Head in this alone*. We cannot be sure what Morales thought the Sudarium was. If he thought the sudarium was the shroud, he would say to wrap the body instead of the head. However, the relic in Oviedo is big enough to wrap a head. In any case, it is certain that he calls both shroud and Sudarium the same relic and interprets the preposition *"de"* as "part of".

The analysis of these documents casts doubts on the presence of the Sudarium of Oviedo inside the Holy Ark. It is more probable that the relic named as *"de sudario"* in some documents and replaced by *"de sindone"* in others (see later), refers rather to the part of the shroud preserved until today in the Cathedral of Oviedo. If this were the case, the Sudarium might not have arrived inside the Holy Ark. The presence of the Sudarium at the end of the list of the Book of Testaments, in addition of the shroud at the beginning, is odd. Although it is just speculation, the bishop could consider that the cloth inside the Holy Ark referred to the Shroud, but reading the Langreo document where part of the Sudarium is mentioned instead, could hesitate and also included the Sudarium at the end of the list.

In the bulletins (sheet for pilgrims) of the 15th century, the list of relics follows the list of the Book of Testaments. The list specifies among the first group the shroud (*de sindone*) but in this case there is no mention of the sudarium in the whole list[62].

The Sudarium is missing in the first printed bulletin of 1493, as well as in the handwritten bulletin of 1535, and was already missing in another document prior to 1493, the *Escritura de los Privilegios de la Cofradía de la Cámara Santa* (year 1465), that is the document of privileges for a brotherhood supporting the Holy Chamber. The lack of reference also happens in the document

---

[60]Sábana

[61]Sudario

[62]López Fernández, Enrique. (2004). *Las Reliquias de San Salvador de Oviedo*. MADU. Granda-Siero. p. 236. López Fernández, Enrique. (2018). *El Santo Sudario de Oviedo*. Soluciones Gráficas. p. 260. López Fernández, Enrique. (2008). *Historia de un Silencio. Sudario de Oviedo a través de los siglos*. MADU. p. 344.

of the foundation of the same brotherhood (year 1344), the Inventory of jewels and relics (years 1305 and 1385), and likewise in the chapter agreements. After several centuries, the first document that refers to the Sudarium is the minutes of the meeting on 17 September 1557.[63] The meeting discussed the exhibition of the Sudarium from the balcony of the Holy Chamber. It means that the relic already had particularly important veneration. It is significant for our discussion that in these minutes, the Sudarium is named holy shroud ("*sancta sabana*"). There is no possibility to refer to the part of the shroud because the exhibition from the balcony is clearly documented for the Sudarium. It is a clear example of mixing the names of the two relics at that time that today we want to differentiate.

Few years later, other documents mentioning the Sudarium are the Chapter Minutes books (records of agreements of canons Chapter) dated 11 October 1566[64]. Another document is not dated but should be of middle 16th century[65]. It is a draft to request privileges and indulgences addressed to the Pope. In this document, the Sudarium takes a relevant role and wins the second place in the list of relics just ahead of the large part of the shroud. The written account gives the enumeration of relics:

> ... great relics that are in it [the cathedral of Oviedo], among others, are the cross that the angels [...] with other pieces of the True Cross. The holy **sudarium** (*sudario*) that our lord had on his most holy face. with another large piece of the **shroud** (*sábana*) in which our Redeemer was also wrapped in the tomb....

As we see, an evidence of the presence and the prominence of the Sudarium granted today in the Oviedo Cathedral dates only

---

[63]López Fernández, Enrique. (2016). *Historical sources about the relics of Oviedo´s cathedral.* Territorio, Sociedad y Poder, 11, 2016. p. 15. The minutes are in the Archives of the Cathedral, *Acuerdos capitulares*, Libro 8, f. 586v). "*Cometieron a los ss.* [] *para que vean cómo se ha de hazer la reja del valcón de la Cámara Santa para mostrar la Sancta Sábana y den horden en ello juntamente con el Administrador".*

[64]López Fernández, Enrique. (2007). *La Veneración del Sudario de Oviedo a través de la Historia.* Proceedings of Oviedo relicario de la Humanidad. Actas del II Congreso sobre el Sudario de Oviedo. Oviedo. p. 361.

[65]López Fernández, Enrique. (2004). *El Santo Sudario de Oviedo.* MADU. Granda-Siero. p. 39.

from the last half of the 16th century[66]. The arrival of the Sudarium to the celebrity must have happened between the year 1535 when the bulletin mentions the part of the shroud (*de sindone*) but does not mention the Sudarium and the year 1557 when exhibitions of the Sudarium from the balcony were established. Since this time, many pilgrims of certain level asked for venerating the Sudarium. There is no doubt about which relic they are referring to, although, in the Chapter Minutes at the beginning, the Sudarium is sometimes called shroud[67]. The full description and the change in the importance given to the Sudarium is evident by testimony of Ambrosio de Morales in the year 1572. His description matches completely with the Sudarium shown today. It is no longer named as "a part of" but as the whole relic.

On the same subject, the pastoral visit of Bishop Diego Aponte de Quiñones took place in 1588. At that time, the Holy Sudarium had already turned into the most precious jewel of all relics and was the object of all honours[68]. The printed bulletins between 1576 and 1595 and the subsequent bulletins include the Sudarium in a similar way as in the referred documents.

> They found a large part of the Lord's **shroud** (*sindone*), in which He lay wrapped in the tomb; his precious **Sudarium** (*sudarium*), stained with his most holy blood, with which his most beautiful face and his most sacred head were wrapped and covered, which is displayed with due reverence three times a year, namely, on the solemn feast of the said holy relics, which is celebrated on March 13, on Good Friday and also on the feast of the Exaltation of the Holy Cross, on September 14 [...]; A big part of the Holy Cross. Eight thorns of the crown, Lord's tunic [...]; of the diapers in which he lay wrapped in the manger...

Unexpectedly, the shroud and the Sudarium appear together in the two first places. While the shroud is "a large part", the

---

[66]López Fernández, Enrique. (2004). *El Santo Sudario de Oviedo*. MADU. Granda-Siero. pp. 43–46. López Fernández, Enrique. 2007. *La Veneración del Sudario de Oviedo a través de la Historia*. Proceedings of Oviedo relicario de la Humanidad. Actas del II Congreso sobre el Sudario de Oviedo. Oviedo.

[67]López Fernández, Enrique. (2008). Historia de un Silencio. Sudario de Oviedo a través de los siglos. MADU. pp. 268–276.

[68]López Fernández, Enrique. (2004). *El Santo Sudario de Oviedo*. MADU. Granda-Siero. p. 39.

Sudarium is not 'a part' but a whole cloth. It has in addition a specific fest.

It is possible that the "part of the sudarium" included among the relics of the Ark was the part of the shroud present until today in the cathedral and the whole Sudarium was another cloth. It is possible that the Sudarium was not inside the Holy Ark.

There is a testimony of the canon Alonso Marañón de Espinosa about the year 1572 who saw some relics outside the Ark and saved in their own reliquary. Among them and in first place was the Sudarium *that our Lord had on his head when he was in the tomb*[69]. Other relics were also in their own reliquary: a small ivory crucifix made by Nicodemus and at its foot a piece of the Wood of the cross, seven thorns of the Lord's crown, one of the 30 coins with which Judas sold the Lord, two bags from the Apostles Saint Peter and Saint Andrew... and so on until listing all the relics that were outside the Ark. Alonso Marañón referred to the true Sudarium that wrapped the head. The part of the shroud was still inside a reliquary box with other 13 relics of Christ until the 18th century[70]. However, the description *that our Lord had on his head when he was in the tomb*, is confusing enough that it does not allow excluding that he considered the Sudarium was the Shroud of the Gospel which covered Christ in the tomb although the whole body.

The Sudarium was in its own reliquary as Ambrosio de Morales established clearly. The way in which the Sudarium has been preserved since the visit of Ambrosio de Morales did not change almost until the second decade of the 21st century. The description provided by Morales matches pretty well with the way the Sudarium has been shown until a few years ago. Only the disposition of the Sudarium was parallel to the floor (horizontal) according to the report of the ecclesiastic and changed to perpendicular to the floor (vertical position) at some point in time, probably around 1761. The horizontal box that had the relic inside

---

[69]López Fernández, Enrique. (2004). *El Santo Sudario de Oviedo*. MADU. Granda-Siero. p. 44. The manuscript is preserved in the Archive of the cathedral. Cámara Santa, legajo I, box 86: Documents referring to the Holy Relics, f. 15r.

[70]Cabal González, Melquíades. (1986). *Cómo enfermaron y murieron los obispos de Oviedo (siglos XVIII al XX)*. Boletín del Instituto de Estudios Asturianos (c. s. i. c.). N.° 119.

could have been substituted during that time by the new one of mahogany[71].

The Sudarium was attached to a wooden frame that was covered with black velvet. We do not know if the original wood frame has been the same until the full remodelling in 2014. The relic was placed on the velvet, attached throughout the wood with silver nails with large round heads (Fig. 2.12). It had two handles on the sides, also made of silver, to support it during transport and during the blessing with it. A red velvet curtain covering the Sudarium hid it from public view (Fig. 2.13 left). Since October 1987 there was a silk white cloth between the red curtain and the Sudarium that protected this last from contact with the velvet curtain. When the Sudarium was exposed, the covering textiles were raised and turned over on the back of the reliquary (Fig. 2.13 right).

Figure 2.12 The layers of the old reliquary of the Sudarium.

---

[71]López Fernández, Enrique. (2004). *El Santo Sudario de Oviedo*. MADU. Granda-Siero. p. 62.

The frame, with the sacred relic, was kept in a wooden drawer carved with gold and blue on the outside, with a crucifix painted on the front. Ambrosio gives the measurement of the box: "*vara y media, y ancha una vara, y en lo alto aun no una quarta*" (about 126 × 84 × 21 cm). They correspond approximately to the measures of the box used until the beginning of the 21st century (Fig. 2.14) except the height that corresponds to the same box but discounting the frontispiece of the new. It is logical if we take into account that the box was maintained in a horizontal position at the time of Ambrosio's visit and it was put vertically in modern times.

Figure 2.13 The cover of the Sudarium (left ©CES)) in a previous reliquary and the covering cloths once turned over on the back (right).

The description discusses then the place taken up by the Sudarium in the Holy Chamber. The rich box with the Holy Sudarium was at a side of the large Cross, commonly said of King D. Pelayo. At the opposite side was the Cross of the Angels. These elements were still in the main places until the beginning of the current century. However, until recently, the drawer of the Sudarium was in vertical position, displaying a life-size black and white photograph (facsimile) of the Sudarium and it was behind the Cross of the Angels (Fig. 2.15).

Figure 2.14 Box containing the reliquary of the Sudarium (© CES).

Figure 2.15 Position of the Sudarium in the last century (© CES).

It is probably that the frame reliquary of the Sudarium in full contact with the cloth was the same as that used until remodelling of the 2014. Moreover, the dimensions of the external box were the same or remarkably similar to those given by Ambrosio, about 126 cm height. It did not fit into the Holy Ark that is 119 cm in its larger external dimension. The Sudarium could not have come inside the Holy Ark with its box. If the Sudarium came inside the Holy Ark, it must have been folded and its particular reliquary must have been manufactured after taking out the Sudarium from the Ark to provide an appropriate container to it.

The Sudarium was out of the Ark at least since the end of the 16th century. We do not know since when it was out of the Ark or even if it was inside the Ark at any time. If the Sudarium did not arrive inside the Ark, the question is: when did it arrive at the cathedral? Well, the same documents that give the Sudarium total prevalence in the second half of the 16th century provide its origin. The structure of these documents starts with the story of the Holy Ark and then, affirms that the Sudarium arrived inside the coffer. However, the presence of the Sudarium inside the Ark is assumed as a fact in the past. It is less reliable than the description of the cloth that is seen directly by the author of the accounts.

With this state of knowledge, we should research more about the possibility of the Sudarium to Oviedo arriving by some other way, some time before the second half of the 16th century. The true documented history of the Sudarium of Oviedo starts around that time. The previous history of the Sudarium is only supposed to be the history of the Ark.

2.7   The History of the Holy Ark

As already mentioned, there are several documents that indicate that the Sudarium was in the Holy Ark for centuries before the opening of the Ark in year 1075. If it were true, the ancient history of the Sudarium would be the history of the Ark with the Sudarium inside. But in all the accounts of the avatars of the Ark previous to the opening at Oviedo there is no express mention of either the Sudarium or almost any of the other relics it contained.

All are included under the generic expression of the Ark of the relics.

According to the donation document, the Ark has its origin in Toledo where some relics were put together inside a plain Ark (*in quadam archa*). This Ark would have arrived in Oviedo later, remaining closed for a long time in the church. A short time later, documents tell us about an older previous reconstructed history which seems to be a mixture of tradition and legend. According to these accounts, the Ark stored the relics inside on its way from Jerusalem to the Holy Chamber.

Another source dated between 1082 and 1096 testifies the transfer of relics from Jerusalem to Oviedo passing through Toledo. It is a letter sent by Osmundus de Astorga to Ida de Boulogne, where he refers to a tradition recorded in some writings, according to which, when the persecution threatened Jerusalem, seven holy men sailed from the Holy city to Spain arriving first at Toledo with the hair of the Virgin and many other relics. When Saracens threatened Spain, the bishops and the believers took every important thing and went to Astorga and Oviedo[72].

The documentation of the old history of the Ark was mainly provided by Mark Guscin.[73] Among other references, he studied the Oviedo's bishop Pelayo's most famous work: The Book of Testaments. It is a collection of documents recording all the donations made to the cathedral church of San Salvador in Oviedo.

The first document recorded by Pelayo was written a few years after 1118[74] and relates the history of the ark. In times of the byzantine emperor Phocas (602–610) the Persians started a war against the empire. When Phocas died, Heraclius (610–641) became the emperor and fought against the Persians and their leader Chosroes. Because of the pagan invasion of Jerusalem (614) and the devastation wrought by Chosroes in the temple of the Lord, a wise decision was taken by the Christians in the Holy City.

---

[72]Henriet, Patrick. (2016). The epistle from Osmundus, bishop of Astorga, to the countess Ida of Boulogne (before 1096). Territorio, Sociedad y Poder, n° 11, 2016. pp. 63–75.

[73]Guscin, Mark. (2006). *La Historia del Sudario de Oviedo*. Ayuntamiento de Oviedo. Avilés. Guscin, Mark. 1999. *Recent Historical Investigations on the Sudarium of Oviedo*. Proceedings of the 1999 Shroud of Turin International Research Conference, Richmond, Virginia.

[74]Rucquoi, Adeline. (2016). *The Manuscript of Cambrai 804: the relics of Oviedo and their miracles*. Territorio, Sociedad y Poder, n° 11, 2016. p. 83.

An Ark that had been made in Jerusalem by the disciples of the apostles with various relics of saints was moved first to Africa by Philip, presbyter of Jerusalem and companion of the presbyter Jerome. It was then taken to Toledo by Fulgentius, the bishop of the African church of Rusp. The relics were kept in the Ark in Toledo, in order to save them from the Islamic advance. The ark remained there from the time of the king Sisebutus (612–621) through various royal successions of Spain, up to the death of king *Don* Rodrigo (710–711), when it was taken to Oviedo. The Ark at this time resided in tents, just like the Ark of the Covenant before the temple was built, up till the reign of Alfonso the younger (791–842), also called the Chaste.

King Alfonso was the first to make Oviedo the capital of the kingdom. In his wise thought he concluded that the above-mentioned Ark should be within the borders of his kingdom, and decided to build a temple for keeping the Holy Ark. The church was devoted to Jesus Christ our Saviour and Redeemer. Since that time, it has been called simply the *San Salvador* (Holy Saviour).

As Guscin says, Pelayo's account coincides with known history in almost all aspects. The order of events is correct: Phocas, Chosroes, Heraclius and the invasion of Jerusalem. The Persians captured Jerusalem in 614, and the Persians took fragments of the true cross. The historian Theophanes[75] states that the Persians took many people into exile along with the true cross. The detail about the fragment of the cross is important because it proves that there were relics in Jerusalem in the times of Persian invasion. If the sudarium was indeed in Jerusalem when Persians attacked the city, it is logical that Christians took the Sudarium out of Jerusalem to save it from the Chosroes' hands.

The Holy Ark, according to these accounts, first reached Africa with the relics from Jerusalem and from there to Toledo in Spain. As will be seen from other sources, it seems unlikely that the Ark was taken directly to Toledo. The Ark left Toledo soon after the Arabs invaded Spain in 711. In any case, the Ark was taken to the Holy Chamber in Oviedo; popular devotion grew steadily from the ninth century on, converting Oviedo into a popular pilgrimage destination.[76]

---

[75]Theophanes, Chronographia 252.
[76]Bennett, Janice. (2001). *Sacred Blood. Sacred Image. The Sudarium of Oviedo*. Libri de Hispania. Colorado.

There are various manuscripts containing all or part of this *corpus pelagianum*. Two manuscripts with versions of the history of the ark and its relics, are in the town of Valenciennes (France). The first of these manuscripts, Valenciennes 99, was written after 1140 and depends on the Pelayo's account.[77] The history in the Valenciennes 99 manuscript is in fact nothing more than a list of cities through which the ark travelled: Jerusalem, Africa, Cartagena, Toledo, and Oviedo. In this case, the sudarium is included in the sixth place of the list of relics together with the shroud (*sindone*) and immediately after it. This is the major justification to think that the part of the shroud and the Sudarium came together in the Ark as two different cloths. However, the expression is still *a part* of the Sudarium ("*de Sudario dni*" (Fig. 2.16). The dependence of the Valenciennes 99 on the Pelayo's list of relics suggests that the author of the French codex found it more correct to reposition the Sudarium from almost the last place, isolated from the other Christ relics, to a place among them.

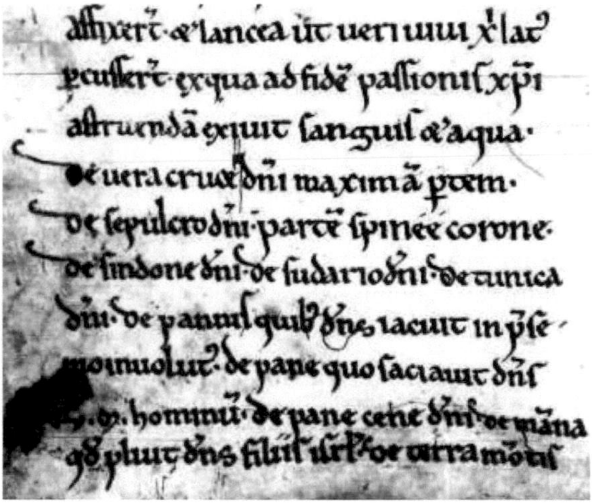

Figure 2.16 Codex 99 Valenciennes 2v (© Bibliothèque municipale Valenciennes).

The other manuscript in the town of Valenciennes is the Valenciennes 30. A third manuscript is the B 804 in the nearby town of Cambrai. And the fourth manuscript is the II 2544 of the

[77]Fernández Conde, Francisco J and Alonso Álvarez, R. (2017). *The catalogs of relics of the cathedral of Oviedo*. Territorio, Sociedad y Poder, 12. p. 69.

Bibliothèque Royale of Brussels. In these documents the version is full of supernatural events and stories of miracles and neither the shroud nor the sudarium is mentioned in the list of relics. Two holy men, called Julianus and Seranus, put all the relics they could in a chest and left it to float away in the sea. The Ark arrived at Carthage. As Guscin remarks, there is no difference in Latin between Carthage in Africa and Cartagena (New Carthage) in Spain. The reference can be then a rumour that the authors heard about the Ark passing by Cartagena in Spain, a Mediterranean port. The two holy men, Julianus and Seranus, died and their bodies remained with the ark that was taken to Toledo when Ildefonso was archbishop (657–667). When the Moslems followed with the invasion of Spain under Tariq, the ark was taken to Asturias where it remained in the mountains (*Monsacro*, or the Holy Mountain) for 45 years.

In this version, some dates are not coherent.[78] According to this version, the ark spent 45 years in the Asturian mountains. As the Ark left Toledo in 711, it would bring us up to 756 for the arrival of the Ark to Oviedo. It is incoherent because the city of Oviedo was not founded until five years later, in 761, when the first church was built, and houses grew up around it. The story in these manuscripts includes more religious fantasy. But we have to remember that it describes alleged events from the past of which the authors were not direct witnesses.

One other source is the Chronicle of the Monk of Silos[79]. A monk from the Spanish monastery of Silos and contemporary of Pelayo of Oviedo wrote the Chronicle. Completely different histories of the Ark show that they did not use the same sources. It says that the Ark went directly from Jerusalem by sea to Seville (Hispalis) in the south of Spain and stayed there for some time before leaving for Toledo. Therefore, a tradition linking the relics with Seville is as old as any other. The Chronicle of Silos says that the Holy Ark remained in Toledo for one hundred years and then travelled from Toledo to Asturias by sea. This is, however,

---

[78]Guscin, Mark. 1999. *Recent Historical Investigations on the Sudarium of Oviedo*. Proceedings of the 1999 Shroud of Turin International Research Conference, Richmond, Virginia. §3. The Manuscripts from North East France and Belgium.

[79]Manuscripts BN 8592 and BN 1181 from the Biblioteca Nacional in Madrid. The Latin transcription of the 28th chapter is in appendix VIII of Guscin, Mark. (2006). *La Historia del Sudario de Oviedo*. Ayuntamiento de Oviedo. Avilés.

totally unbelievable. Toledo is in the middle of the peninsula and far from any maritime port. It would be necessary to travel more distance to a port than directly to Oviedo. The lapse of time in Toledo (100 years) is also problematic. In this reference, there is no list of relics.

Another source of the history of the Ark is the *Chronicon Mundi* of Lucas of Tuy. He travelled by many countries that embraced Christianity and was named the bishop of Tuy in 1239. The date of the composition of his work must be after 1236, years after the Silos and the other mentioned sources of the Ark history. It gives the following information:[80]

> "When the terror of the Gentiles was pressing, in the time of Mohammad, the false prophet, the ark was taken by boat from Jerusalem to Seville. Then it stayed in Toledo for 95 years, and when Toledo was attacked by the Moors, Pelayo took the ark of God and took it to Asturias through hidden places. As has already been said, King Alfonso put it in the church at Oviedo with much honour and the relics of many other saints".

Lucas has improved the Chronicle of Silos. He omits the ridiculous detail of the trip from Toledo to Asturias by sea. As Guscin discovered, the more reliable manuscripts[81] of the Lucas work reads 75 years and neither 100 nor 95 years. With this correction the version of Lucas gains credibility. If the Ark was really in Toledo for 75 years, and it left this city in 711, then this gives us 636 for the moving. It was the perfect time for the Ark and its relics to be taken from Seville to Toledo.

Taking in account all these references, the history of the Holy Ark can be summarized as following:[82]

When the king of the Persians, Chosroes II, invaded Jerusalem in 614, some Christians fled with the Holy Ark and some relics.

---

[80]Andreas Schott. (1608). *Hispania Illustrata*, Vol. IV. page 74 translated by Guscin, Mark. (1999). *Recent Historical Investigations on the Sudarium of Oviedo*. Proceedings of the 1999 Shroud of Turin International Research Conference, Richmond, Virginia. §4. Lucas of Tuy.

[81]Folio 58r of the thirteenth century manuscript at Salamanca (2248) and the unnumbered folio of Seville 58-4-43, Guscin, Mark. (1999). *Recent Historical Investigations on the Sudarium of Oviedo*. Proceedings of the 1999 Shroud of Turin International Research Conference, Richmond, Virginia. §4. Lucas of Tuy.

[82]Guscin, Mark. (1999). *Recent Historical Investigations on the Sudarium of Oviedo*. Proceedings of the 1999 Shroud of Turin International Research Conference, Richmond, Virginia.

The flight was justifiable, because Chosroes was searching for relics like the Holy Cross, among other things. The ark left Jerusalem by sea, with a possible stop in a city on the northern coast of Africa (most probably Alexandria). The Ark could enter Spain via Cartagena, and then go to Seville that was under the bishop Isidoro. When he died in 636, the Ark was moved to Toledo. The ark stayed in that city for 75 years, until the invasion of the peninsula by Tariq in 711. By that time, Christians carried the chest to the safer north of the country. The Ark was then hidden for approximately half a century in the mountains of Asturias till the Arabs were defeated in this region.

The existence of an Ark in Toledo is also recorded in the Third Council of Braga[83] (*arcam Dei cum reliquiis*) celebrated in 675. It is a contemporary document of the event. The transfer of relics by Christians to the mountains of Asturias is also endorsed by a document, relatively contemporary to the event, written by the Arab author Albunbenque Mohamat Rasis[84] who wrote it around the year 977.

At present, the Holy Ark is a black oak wooden chest, partially covered with silver gilt sheet. It has a prismatic shape, and its measurements are: 1.19 m long, 0.93 m wide and 0.82 m high, including the top and the stringers. The lower board is missing, which perhaps never existed or may have deteriorated due to humidity since an imprecise date. It suffered a moth on only one of its boards[85]. The five silver plates that cover the boards are decorated with different scenes from the life of Christ.

After the blowing up of the Holy Chamber in 1934 (Fig. 2.17), the Don Juan Institute of Valencia in Madrid restored the Holy Ark. They found 13 coins trapped between the silver foil and the wood. They belonged to kings such as Sancho IV, Fernando IV or Alfonso XI, all of them from the 13th century onwards. In the table on the left side, there was a trace of an arch poly-lobed engraved with the point of a compass and a peculiar punch typical of Andalusia. It has been dated between the end of the

---

[83]Guscin, Mark. (2006). *La Historia del Sudario de Oviedo*. Ayuntamiento de Oviedo. Oviedo Avilés 2006. pp. 175–176. Canon VI of the Council of Braga. Latin transcription in the appendix XI of the reference.

[84]*Historia y Descripción de España* cited by Guscin, Mark. (2006). *La Historia del Sudario de Oviedo*. Ayuntamiento de Oviedo. Oviedo Avilés 2006. p. 170.

[85]Gómez Moreno, Manuel. (1945). *El Arca Santa de Oviedo Documentada*. Archivo español de arte. 18. N.º 69. pp. 125–136.

10th century and the beginning of the 11th. The oldest age of this Ark could be then of few years before the opening[86]. However, it seems probable that it was commissioned by King Alfonso VI. About the older Ark, coming from the Holy Land according to legend, it does not seem to have been preserved. The lost of the original Ark is surprising because it was considered at that time as executed by the disciples of the apostles and we would expect that it would have been kept as one more relic.

Figure 2.17 View of the debris after the blowing of the Holy Chamber in 1934, once performed the clearing up. The cover of the Holy Ark under the Cross of the Angels (left) and the front of the Holy Ark (right) (© Boletín de la Academia de la Historia. Friday 9 November 1934).

The report of the *Academia de la Historia* on Friday 9 November 1934 about the blowing of the Holy Chamber[87] says that the large box or drawer containing the Holy Sudarium slid over the rubble remaining perfectly unharmed and the Holy Sudarium appeared intact inside its box.

---

[86]Barta, C. (2007). *Datación del Sudario de Oviedo*. Proceedings of Oviedo relicario de la Humanidad. Actas del II Congreso sobre el Sudario de Oviedo. Oviedo.

[87]Report M. Gómez Moreno (Acta Sesión 09/11/1934). Boletín R AH, 105.

2.8   Modern Scientific Interest

We can say that the scientific study of the Sudarium of Oviedo started in August 1965 when Msgr. Giulio Ricci (Fig. 2.18) visited the Cathedral of Oviedo and looked at the relic as an archaeological document. He was a member of the Vatican curia and President of the *Centro Romano di Sindonología*. He was the pioneer in the study of the cloth. He was looking for a complementary cloth of the Shroud of Turin and, at first sight; he thought he had found it. In the second and further editions of his book "*L'Uomo della Sindone é Gesù*", he exhibits the scoop within his work about the Oviedo cloth. He proposed for the first time that the Sudarium had been used to cover the face of Jesus from Golgotha to the tomb and, once there, it was placed separately.

Figure 2.18 Msgr. Ricci (2nd from the left) with the Sudarium of Oviedo (© CES).

He found the symmetry between the two main blood stains of cloth and presented a macroscopically acceptable fit between the Sudarium and the face on the Shroud of Turin. In this overlay, Ricci marks a series of elements that captured his attention.

The first of them was a typical blood stain present on the right side of the mouth of the man from the Shroud of Turin that he observed also in the Sudarium of Oviedo.

The second element that drew Ricci's attention was the matching of the beard by the perfect superposition of the face of the Shroud with the life-size photo of the Sudarium.

In August 1977 he visited Oviedo again with an archaeologist and a photographer and took some samples: threads and three small pieces. Msgr. Ricci promoted the visit of Max Frei two years later. Frei was a palynologist (expert in pollen analysis) who had studied the pollen of the Shroud of Turin. He also took dust samples with the method of the adhesive tapes. Ricci was also the intermediary for the visit in May 1985 of Dr Pier Luigi Baima Bollone, the Italian haematologist who had also studied the Shroud of Turin. He applied adhesive tapes, cut seven threads from the bloodiest stains and took a lot of photographs. In 1987, the Sudarium had a place in the Congress of Sindonology held in Syracuse because Bollone presented a communication with Franca Pastore Trosello, on the weaving of the Sudarium. In the second International Symposium for the Shroud of Turin held in Rome in June 1993, Dr. Carlo Goldoni presented his result about the blood group of the stains of the Sudarium that was AB.

In 1989, the *Equipo de Investigación del Centro Español de Sindonología* (EDICES. Research Team of the Spanish Centre for Sindonology) was formed to continue the investigations of the Sudarium. Its first president was D. Guillermo Heras, the third from the right in Fig. 2.19.

Figure 2.19 EDICES 1989 (© CES).

The Archbishop, Don Gabino Díaz Merchán, with the consensus of the Chapter of canons of the cathedral gave the permission for the complete study of the Holy Sudarium. His words were "Here, from immemorial times, people have been blessed with the Sudarium on Good Friday and Good Friday is a very serious date, so I want to know what I am doing".

Under the direction of Heras, the research team grew up to more than thirty members (Fig. 2.20 and Fig. 2.21). They came from Spain and abroad. To mention some of them, we can name G. Heras, J.M. Rodríguez, D. Villalaín, F. Montero, J. Izquierdo, M. Carreira, S. Mantilla, T. Ramos, J. García Iglesias, R. Sánchez Crespo, F. Díez, E. Monte, R. Platero, F. Pendás, J.M. Miñarro, A. Muñoz-Cobo, A. Ordoñez, M. Guscin, J&R. Jackson, K. Propp, A&M. Whanger, J. Bennet, etc. Other important collaborators were B. Bollone, C. Goldoni, B. Barberis, N. Ballosino, etc. Some of the members have passed away since. C. Barta, the author of this book, joined the team at the beginning of the year 2000 with the main mission of analysing radiocarbon dating. He stayed in the team for eighteen years until the new director Alfonso Sánchez Hermosilla reduced the team to nine Spanish members. It is worth mentioning that the interest and collaboration of the first bishop Mons. Diaz Merchan and his successor D. Carlos Osoro and also of the Chapter of canons have been the key factors for the success of the research.

The results of the EDICES research have been presented in two congresses organized by this group. In the "I International Congress on the Sudarium of Oviedo", held from the 29th of October to 1st November in 1994, the preliminary conclusions of the research were presented. There were six sessions for 26 communications. Most of them were presented by the Spanish members of the EDICES. Among the foreign speakers were J&R Jackson, P.L. Baima Bollone, A. Whanger and C. Goldoni.

The main novelties presented in the first congress were in first place the sequence of the use of the Sudarium from the cross to the tomb with estimated intervals of time and position of the corpse in each phase. It was the result of the work of the team led by Dr. Villalaín and G. Heras (Fig. 2.22). The other important progress was the microscopic physical and chemical analysis carried out by F. Montero. The 3D reconstruction of the head of the man of the Sudarium performed by A. del Campo, the hematologic study presented by C. Goldoni and A. Adler, the superposition of the stains of the Sudarium on the face of the Shroud provided by A. Whanger and the first radiocarbon dating communicated by the Italian team through Baima Bollone as the speaker is also worth mentioning. The proceedings of this congress were made available for the scientific world[88].

Figure 2.20 EDICES 1994 (© CES).

Figure 2.21 EDICES 2006 (© CES).

---

[88]AA.VV. (1994). *Sudario del Señor*. Proceedings of the first International Congress about the Sudarium of Oviedo. Universidad de Oviedo. Oviedo.

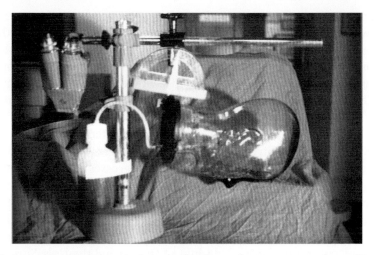

Figure 2.22 Experiment for verification of sequence and position performed by Dr Villalaín. (© J. M. Rodríguez Almenar & CES)

In the "II International Congress on the Sudarium of Oviedo" held from the 13th to the 15th of April in 2007, a series of aspects proposed in the previous congress were confirmed as explanation and meaning had been found in later years. Many initial conclusions were verified, and new ones appeared, but in both cases, we speak of results that were subjected to prior internal criticism and that were published only after long discussion.

The newer results were the DNA of the blood of the Sudarium, the documentation of the history of the relic, the matching between Sudarium and Shroud of Turin, the 3D reconstruction of the head and whole body wrapped in both cloths, the Sudarium and the Shroud. Moreover, some advice was provided for the conservation of the cloth at the cathedral. Simultaneous translation was also provided, which was another example of confusion between the words *sindon* and *sudarium* when the translator referred to the Shroud of Turin as the Sudarium[89]. This confusion happens today in English as it happened in the past in Latin, as we previously mentioned.

César Barta, the author of this book, was entrusted with one of the presentations about the dating of the Sudarium and with

[89]Guscin. Mark. (2007). The Second International Conference on the Sudarium of Oviedo. Shroud Newsletter, N.65.

one of the presentations dedicated to the matching between the two purported funeral cloths of Christ. While the talk on the radiocarbon dating was being delivered, the Archbishop D. Carlos Osoro listened in the first row. He was a promoter of this second congress showing a broad minded attitude for scientific studies. The issue of radiocarbon dating is still open. The proceedings of this congress were also made available for the scientific world[90].

In addition to the two congresses promoted by the EDICES, there are other contributions of the team in international meetings. Before the organization of the two congresses in Oviedo, the first intervention at an international level by members of the research team was at the Sindonology Congress held in Cagliari in April 1990. In the proceedings, the Spanish presentation entitled *"El Sudario de Oviedo y la S. Sindone: Dos reliquias complementarias?"* (The Sudarium of Oviedo and the Holy Shroud: Two complementary relics?) is a 70-page document (35 in Spanish and another 35 in Italian) with large size photographs. G. Heras remarked the coincidence between the facial bloodstains on the Shroud of Turin and on the cloth of Oviedo. This results cast doubt on the C14 date attributed to the Shroud. This communication also triggered the knowledge of the Sudarium to the world.

Among other attendances of the members of EDICES in international meetings to spread the knowledge of the Sudarium, one happened at the Nice Congress in 1997, where Mark Guscin was the spokesperson for the team. Also, the International Congress "Shroud 97", organized in the Republic of San Marino attended by J.M. Rodríguez Almenar. Another monograph was presented in the "III International Congress of Sindonology" held in Turin in June 1998. Two other congresses that were attended by members of EDICES were the 1999 Richmond Congress and the Orvieto World Congress in 2000. In Richmond, Mark Guscin showed a live demonstration of how the cloth would have wrapped the head. The volunteer was Barrie Schwortz (Fig. 2.23).

In the Orvieto Congress, the first communication of C. Barta was presented even before becoming a member of EDICES. It addressed the *Sindone* sample from Constantinople in Toledo (Spain). The content of that presentation has been developed in paragraph 1.3.3 Sainte Chapelle of Paris in the first chapter of this

---

[90]AA.VV. (2007). *Oviedo relicario de la Humanidad.* Proceedings of the second International Congress about the Sudarium of Oviedo. Oviedo.

book. Moreover, in the same congress, the Italian team of M. Moroni et al., for the first time suggested an explanation of the C14 dating results for both cloths, the Shroud and the Sudarium, based on the neutron irradiation[91].

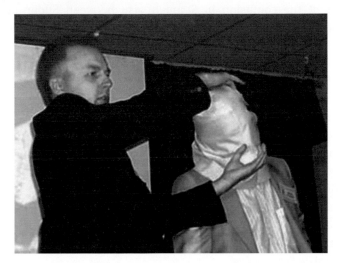

Figure 2.23 Mark Guscin (left) and Barrie Schwortz (under wraps) in the Richmond Congress (© Barrie Schwortz. Shroud.com).

The lectures about the Sudarium continued in further congresses: Dallas 2001, Dallas 2005, Ohio 2008, Frascati ENEA 2010, Fatima 2010, Panama 2012, Bari 2014, St. Louis, Missouri 2014, Turin 2015, Pasco 2017, Jalsa Salana Convention 2017 and Ancaster, Ontario 2019.

A special event worth mentioning is the First International Congress on the Holy Shroud in Spain held in Valencia between April 28–30, 2012. It was organized by the *Centro Español de Sindonología* (CES Spanish Centre of Sindonology) to commemorate the 25th anniversary of the CES creation. There were five sessions for 23 communications. In this case, 40 per cent of speakers were foreigners. Although the main subject of the Congress was the Shroud of Turin, the Sudarium of Oviedo was discussed in more than three communications. The description and the C14 dating of

---

[91]Moroni, M., Barbesino, M., Bettinelli, F. (2000). *A suggestive hypothesis covering the radiodating results of the Sudarium of Oviedo and the Shroud of Turin.* The Orvieto Worldwide Conference "Sindone 2000". August 27–28–29, 2000.

the Sudarium were analysed by F. Montero and the commonalities between Shroud and Sudarium by A. Sánchez Hermosilla.

In the C. Barta's communication "What the Shroud is and is not" the present author also dealt partially with the Sudarium to highlight the confusion in Spanish also between Sudarium and Shroud. As we have pointed out in the case of Latin and English, similar confusion is present in Spanish also[92]. In Spanish America people often use "sudario" for the Shroud of Turin. Therefore, this fact should make us prudent in identifying or interpreting the funeral cloth mentioned as "sudario", "shroud", "suaire" in Latin, Spanish, English, and French.

The other relationship between Shroud of Turin and Sudarium of Oviedo that Barta addressed was the so-called "ponytail" of the Man of the Shroud. It could have been the result of the use of the Sudarium by the Man of the Shroud.

The congress was closed by the Archbishop of Valencia, Carlos Osoro, the same that promoted the second congress in Oviedo. He has been by chance a "follower" of research of the *Centro Español de Syndonología*.

In the Bibliography, the reader will find the published works of the members of the EDICES, as a single author and as co-authors.

2.9 Summary of the Early History of the Sudarium

The Sudarium of Christ could have had a knot as Nonnus de Panopolis said. It could have been preserved on the shore of the Jordan River. Christians in Jerusalem could have put it in the Holy Ark around the year 614 and the Sudarium of Christ could have arrived in Oviedo inside the Ark. However, these steps must be verified as far as possible through the scientific research that EDICES and other academics initiated in the past and continue till today.

---

[92]Before the discovery of the Sudarium of Oviedo by Ricci, a series of books edited by the *Biblioteca Sindoniana* in Spanish (in 1957) called the Shroud *santo sudario*.

Chapter 3

Scientific Analyses

In this chapter, we describe the scientific analysis included in the specific congresses of the Sudarium of Oviedo held in 1994 and 2007 and elsewhere. The tests include but are not limited to photography infrared, ultraviolet, transparency, scanning electron microscopy–energy dispersive X-ray (SEM-EDX), textile material, dust, pollen, aloe, myrrh, folds, holes, stitching, etc.

3.1 Description of the Sudarium of Oviedo

After the previous "discovery" by Monsignor Ricci and other Italians recommended by him, the direct analysis of the Sudarium has been performed almost exclusively by the EDICES (Research Team of Spanish Centre of Sindonology) with the agreement of the Council of Canons. The scientific observation of the Sudarium of Oviedo was carried out particularly by Felipe Montero[1].

According to Morales (see Chapter 2), in 1572, a wooden frame covered with black velvet provided support for the Sudarium which was fastened to it with a silk cloth. The frame had two

---

[1]Technical Chemical Engineer, Deputy Director of the E.D.I.C.E.S., Ex-president of the *Biodeterioro y Biodegradación de la Sociedad Española de Microbiología*.

The Sudarium of Oviedo: Signs of Jesus Christ's Death
César Barta
Copyright © 2022 Jenny Stanford Publishing Pte. Ltd.
ISBN 978-981-4968-13-3 (Hardcover), 978-1-003-27752-1 (eBook)
www.jennystanford.com

silver handles[2]. The assembly of the Sudarium in its frame underwent several changes along the centuries. Until the remodelling of the 2014, the linen appeared mounted on an old silvered wooden frame. The silver frame, which covers the wooden one, was made up of six fragments supported on a frame of wood by means of silver screws. The unions of the six fragments by means of these screws were decorated by golden reliefs. This silver frame was a gift (on 20 October 1763) from D. Juan Francisco Manrique de Lara Brabo y Guzman, Bishop of Plasencia, who also had been Bishop of Oviedo from 1754 to 1760. Although the frame was the same for years, the inner fabrics that supported the Sudarium changed.

Before the visit of Mons Ricci in 1977, the Sudarium showed the side that was in contact with the skin of the corpse. This side is more polluted than the opposite side. We do not have a picture close-up of that arrangement, yet it can be recognized with the pictures of those years. We can notice at the right lower corner in Fig. 3.1 that the cloth is still complete, and no piece has yet been removed. In the visit of 1977, the council of canons gave to Ricci a support cloth that has been in contact with the Sudarium probably since the 16th century. At that moment, the back side of the Sudarium was examined. It was cleaner than the displayed side during centuries[3].

In the visit of 1977 or during one of the other interventions for cutting samples for Ricci, the Sudarium had been turned to show the opposite cleaner side. This probably happened in 1986, when Dr. Luis G. Saavedra on 3 April took a piece under request of Mons Ricci for the hematologic and dating analysis. Figure 3.2 shows the Sudarium as it looked before October 1987. In this case, it showed the side that was not in contact with the skin of the corpse. All around the perimeter remains the previous silver ribbon. It did not hide the spot from where the piece had been taken for tests. At the low right corner, we can notice the place where the piece that was probably taken by Dr Luis G. Saavedra 3 April 1986 was.

[2]López Fernández, Enrique. (2018). *El Santo Sudario de Oviedo*. Soluciones Gráficas. p. 53.
[3]Ricci, G. (1992). *La Sindone contestata, difesa, spiegata*, Collana Emmaus, Roma p. 267.

Figure 3.1 Sudarium of Oviedo during a procession through the streets of Oviedo (1942). (© E. López-Soluciones Gráficas).

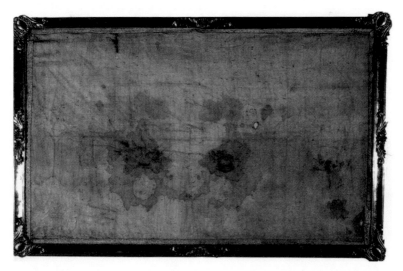

Figure 3.2 Sudarium of Oviedo as it appeared before October 1987 (© CES).

In October 1987, the Benedictine nuns of St Pelayo repaired the assembly[4]. The Sudarium was dismounted and detached from two white cloths that were in turn nailed to the wooden frame.

---

[4]*"Santo Sudario"* informe Monjas Benedictinas, segunda quincena de octubre 1987.

The nails that were removed had rust, which affected some parts of the Holy Sudarium at the edges. The silver ribbon that outlined the perimeter was also detached. It served to cover the nails and the white cloth edge that was visible, since the Sudarium was smaller than the wooden frame on which it was placed.

The Sudarium was then sewn on a new linen fabric. The nuns folded approximately 1 cm along the edge that had the cutting in the corner to disguise the scissor cut. After cleaning the wooden frame, the new support linen was stretched with a cotton tape. By this way, the Sudarium was stretched, without the tension affecting it directly, thus avoiding the use of nails that could damage it. The nuns put a natural silk curtain in between the Sudarium and the red embroidered curtain and attached to the latter to prevent it to rub the Sudarium. The thread used was also made of natural silk. After the repairing, the visible side of the Sudarium was still the one that was not in contact with the skin of the corpse.

Figure 3.3 shows the Sudarium as it appeared after the repairing of the nuns and when the members of EDICES first saw it on 10 November 1989. There was no ribbon around the perimeter and the cut in the corner was concealed with a fold. The visible side was still the one that was not in contact with the corpse.

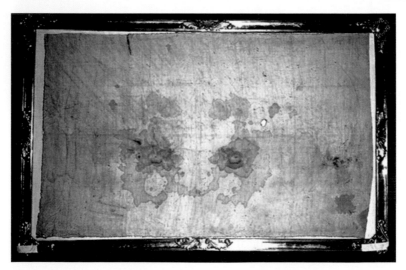

Figure 3.3 Sudarium of Oviedo as it appeared in 1989 (© CES).

In this chapter, we present a revised version of the description of the Sudarium presented by members of the EDICES in the three congresses promoted by the CES[5].

The cleaner side remained exposed to the visitors until 25 November 2006, when it was turned around again following the advice of EDICES. It was then displayed without the ribbon and without disguising folds. The cloth then showed the side that was in contact with the skin of the corpse (Fig. 3.4).

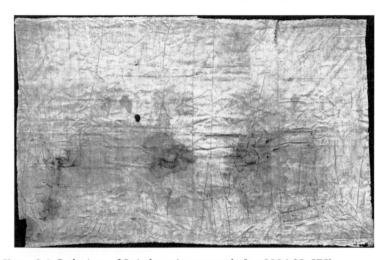

Figure 3.4 Sudarium of Oviedo as it appeared after 2006 (© CES).

It should be noted that the correct position for the observation of the Sudarium is in landscape format, that is, with the shorter dimensions of the cloth along the vertical direction and a seam of 5.5 cm that is observed on the upper edge, remaining slightly on the right part of the rectangle. This will be named for reference the **standard arrangement**.

The conservation and presentation of the Sudarium of Oviedo changed in April 2014 after several years of the project's development. In the 2nd International Congress for the Sudarium of Oviedo in 2007, members of EDICES advised the conditions under which it should be stored and displayed[6]. However, the

[5]Montero, Felipe. (2002). *Description of the Sudarium of Oviedo.* Proceeding of the 1st International Congress for the Shroud. Valencia 28–30 de April 2012.

[6]Mantilla, S., and Montero, F. (2007). *Protección y conservación del Santo Sudario de Oviedo.* Proceedings of the 2nd International Congress for the Sudarium of Oviedo. pp. 95–100.

agreement with the canons Chapter and the need for financing extended the completion of the project for 7 years.

Finally, the Sudarium is preserved in accordance with almost all of the EDICES recommendations. It is stored supported by a rigid polycarbonate support in an inclined position of 37 degrees to reduce the stress of its own weight and without any other canvas covering it. It is currently in a sealed container with a nitrogen atmosphere to prevent the growth of microorganisms. This sealed urn or reliquary protects the Sudarium from variations of temperature and relative humidity. In order to venerate it, the upper part of the reliquary is made of anti-glare laminated safety glass (Fig. 3.5). The contour is decorated with the silver frame recovered from the ancient reliquary.

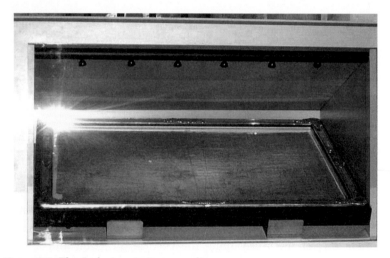

Figure 3.5 The Sudarium in its new reliquary.

But the new reliquary is enclosed in another larger container that protects it from light since it is opaque on all sides except the front (Fig. 3.6). What visitors see is a facsimile that covers the window of this second container (Fig. 3.7). The temperature and relative humidity inside this external container are maintained at around 18°C and 50±3% respectively by a climatic control system[7].

---

[7]López Fernández, Enrique. (2018). *El Santo Sudario de Oviedo*. Soluciones Gráficas. p. 183–184.

Figure 3.6 The Holy Chamber restored after 2014 (© Jorge Hevia).

Figure 3.7 The facsimile of the Sudarium that covers the window of the primary reliquary container.

If interested people come to contemplate it, it is exceptionally, illuminated with cold light, using fibre optics. As tradition dictates, the Sudarium is venerated three times a year: Good Friday, September 14 and 21. For these events, it is taken out of the container, but it is still in its sealed reliquary and on a support that maintains the 20 kg of the set. This is how it is presented in front of the altar for the worshippers to approach. The bishop can no longer lift it by directing it towards each of the cardinal directions to bless the whole world.

3.2   Photography—Infrared, Ultraviolet, and Transparency

Several types of photography have been applied to the Sudarium observation:

- Conventional photography
- Infrared reflection photography
- Ultraviolet light photography
- Photography for transparency
- Photography lighting from a side (grazing light incidence)
- Electronic image processing

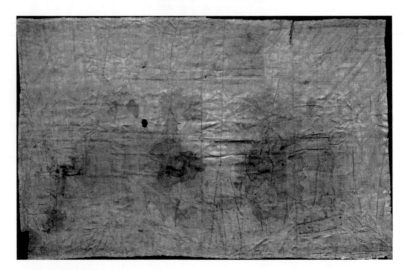

Figure 3.8 Visible light photography of the Sudarium (© Website of the Cathedral of Oviedo).

The visible light photography is the obligatory reference for all the others to present the cloth in all its extension. Figure 3.8 displays the visible light photography of the Sudarium out of the reliquary.

The first thing to note is the two almost symmetrical stains at both sides of the vertical axis. They are of brown colour with diverse intensities. We will see later that in the hematologic studies carried out by the forensic team of the EDICES, these stains turn out to be human blood of AB group. More precisely, the central symmetrical stains are a mixture of pulmonary oedema and blood. The smaller stains near the left side are produced when the person was still alive (see Section 4.4.3). They are not symmetrical, and the wounds were originated by sharp objects.

To place the taken samples and to have an accurate location of the features, the team placed over the Sudarium an orthogonal squared grid, in which every unit corresponded to 1 cm (Fig. 3.9). There are other coordinates systems used during the research[8].

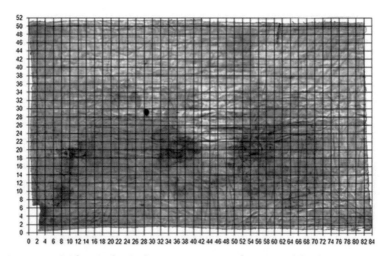

Figure 3.9 Grid over the Sudarium to accurate location of the features.

[8]Montero, F., et al. (1998). *El Sudario de Oviedo. Descripción del Lienzo.* El Santo Sudario de Oviedo. Hallazgos Recientes. CES. Valencia. p. 253/37. In this case, the coordinates are numbers (1–34) and letters (A-U) for squares of two and a half centimetres of side, although it written only two cm. Also Heras, G. (1994). *Descripción General del Sudario de Oviedo.* Sudario del Señor. Proc. 1st International Congress on the Sudarium of Oviedo. Universidad de Oviedo. Oviedo. p. 50. The coordinates are numbers (1–42) and letters (A–Z) for squares of two centimetres of side.

For the following sections, we will give numbers for the wavelength of the light and we must define the correspondence between these values and the infrared, visible and ultraviolet ranges. The visible spectrum runs from about 400 nm (violet) to about 700 nm (red). Values below 380 nm are ultraviolet and values above 750 nm are infrared (Fig. 3.10).

Figure 3.10 Spectrum from the ultraviolet to the infrared.

3.2.1 Infrared Reflection Photography

The infrared inspection was made with a Kodak High Speed film sensitive up to 900 nm and with a CCD video camera sensitive up to 1100 nm. The illumination was carried out with tungsten lamps filtering the visible frequencies by means of a filter that cuts at 690 nm.

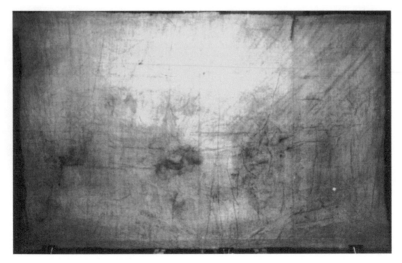

Figure 3.11 Infrared photography of the entire Sudarium (© CES).

Therefore, the recorded images captured the near infrared from 1100 nm to 690 nm (at the limit of the visible, Fig. 3.11). Through this technique some hidden marks at first glance can be detected even if they were under stains. However, in the Sudarium nothing was detected in any area of either of the two surfaces studied. It also allows differentiating areas stained with different concentrations of blood.

It should be remarked that some stains at the lower part of the Sudarium both below the central stain and below the nape stain, have the same properties, showing very poor infrared reflection. This fact supports the hypothesis that these stains have the same origin. That is, the "butterfly" stain below the group of wounds produced by sharp objects could come from the mouth because both stains are similar under infrared light.

3.2.2 Ultraviolet Fluorescence Photography

Before the arrival of the EDICES, on August 31, 1977, Mons Ricci was assisted by Antonio Solazzi as official photographer[9]. He took the best ultraviolet photography. He used two lamps Wood 290 nm with a Kodak Wratten B2 filter for exposure times from 20 seconds to 3 minutes in total darkness. In this way, the fluorescence of visible light excited by ultraviolet (UV) frequencies is recorded. The Sudarium was supported by some threads to the frame once removed all the protection cloths. The largest size of the Sudarium was in horizontal orientation and the picture was taken from the side that was not in contact with the head skin (Fig. 3.12).

The EDICES team[10] also took its own ultraviolet photographs (Fig. 3.13). A near ultraviolet illumination with a mercury vapour lamp and filtering frequencies above 450 nm was used. The image capture was taken with standard colour film and CCD camera.

The images help to highlight the different responses of blood stains to the UV light. These images allow comparing the signal in the wounds of the nape and in the central stains. However, the possible effect of the heterogeneous intensity of the UV

---

[9]Falcinelli, R. (2007). *Análisis sobre la tela de protección del Sudario de Oviedo. Cronohistoria de una investigación.* Proceedings of the 2nd International Congress for the Sudarium of Oviedo. pp. 635–650.

[10]Heras, G. (1994). *Descripción General del Sudario de Oviedo* and Izquierdo, J (1994) *Métodos de Estudio Empleados. Técnicas de Imagen.* Proceeding of the 1st International Congress for the Sudario de Oviedo. Oviedo 29–31 October 1994.

exposition cannot be rule out. In Fig. 3.13 it looks like the lighting of the two UV lamps is greater in the middle and in the right side than in the left side and in the corners. When assessing if there is a different response for different type of blood, vital in the nape and post-mortem around the mouth, we need the same condition of UV exposure and recording in both cases. In Fig. 3.12 of UV photograph, the fluorescence of the clean area near the stains of the mouth and the nape are remarkably close to each other and we can consider that fluorescence in the wounds of the nape is lightly darker than in the mouth.

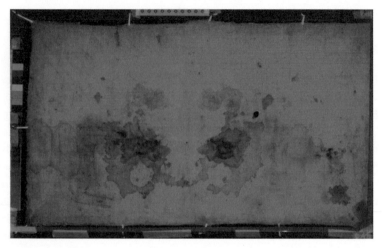

Figure 3.12 Solazzi's Ultraviolet photograph (Courtesy: Roberto Falcinelli from Archive Diocesan Center of Syndonology Giulio Ricci-Rom).

The image (Fig. 3.13) revealed substances that otherwise would go unnoticed. Wax remains have been detected at the edges of a hole present in the Sudarium, as well as other drops of the same nature at other points on its surface. Modern fibre yarn different from the original was also observed under this UV light. It corresponds to the thread used to seam the tear of 5.5 cm being in the top part of the cloth.

With this type of illumination, the contour of the stains is easier to identify. For example, this is the case of the stains corresponding to the wounds originated by pointed objects (thorns?) that have a lighter halo around the darker centre (Fig. 3.14).

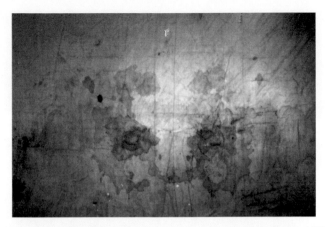

Figure 3.13 Ultraviolet photography of almost the entire Sudarium (© CES).

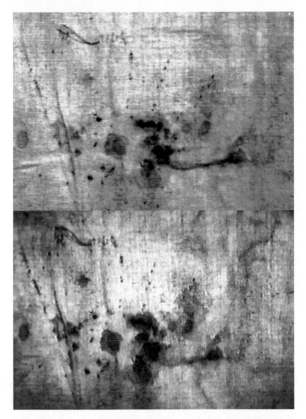

Figure 3.14 Visible (up) and ultraviolet (down) photography of the pointed stains (© CES).

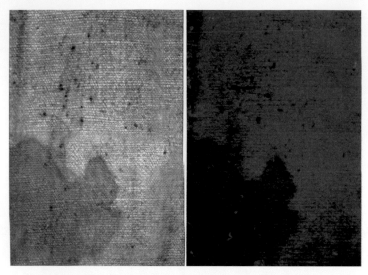

Figure 3.15 Reflected light photograph (Left) and Ultraviolet fluorescent photograph (Right) of the same area (@ Turin Shroud Center of Colorado).

The team of Dr Jackson also took UV images in the inspection of November 2006. They prepared a specific *x-y* positioning rail system over the Sudarium to move the camera together with two fluorescent light sources. Sequential pairs of reflected light and ultraviolet fluorescence light photographs of selected regions of the Sudarium were taken. The reflected light photograph was first taken over a captured size of 3 × 5 cm. A second picture was taken with a 13 second time exposure using an UV light source with all other lights turned off. In this case the UV emission was almost monochromatic because it went through a narrow band pass interference filter that allowed wavelengths only in the range 365±5 nm. The fluorescence in the visible range was captured with the camera after being eliminated the ultraviolet waves under 380 nm. An example of the pair of visible and UV photographs is given in Fig. 3.15 for of a blood stain (part of the "butterfly" stain) on the Sudarium. In this case, there is not a halo at the edge of the stain, and it differs from the pointed stains (Fig. 3.14). This may be due to the difference between the vital nature of the dot shaped stains having coagulated and the post-mortem nature of the "butterfly" stain which is a mixture of blood and oedema.

In the middle of this last blood stain, there are several small fluorescent lint fibrils[11] seen thanks to the brightness of their UV fluorescence. Similar fluorescent lint fibrils were seen in many of the UV fluorescent microphotographs and are thought to be external contaminants over the Sudarium.

3.2.3   Transparency Photography

By lighting from one side and photographing on the other we obtain the transparency photograph. It helps to identify the most convenient spots for taking samples for haematological analysis and for other natures. At the same time, it is possible to differentiate areas of the fabric with different texture and even lack of homogeneity, which helps to the study of the fabric that will be later conducted. The existing perforations in different parts of the cloth are also clearer.

It is also possible to highlight the zones with different amounts of blood (Fig. 3.16), which allows estimating which of the areas have retained the most of it.

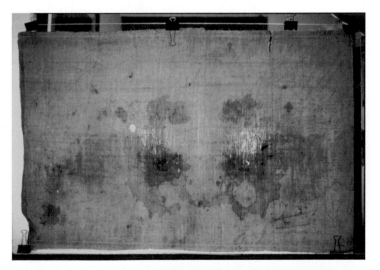

Figure 3.16 Photograph of the Sudarium illuminated by Transparency (© CES).

---

[11]Although Dr. John Jackson wrote "fibrils" in his report, he could refer rather to fibres that are between 3 mm and 80 mm in length. A linen fibre is made up of thousands of fibrils that are up to 3 µm long.

This study helps to determine which one of the stains was the first in contact with the face and to establish the sequence of their formation. With the EDICES photograph, the layer which is not in contact with the face (right) is darker than the layer in contact with the face (left) against the expected. However, it can again be a consequence of the different intensity of the illumination.

Thankfully, we also have the transparency photograph taken by Antonio Solazzi mentioned previously. He also used two lamps[12] but the result look more homogeneous. We note that Solazzi's picture was obtained from the opposite side of the EDICES picture. When the intensity of the image along the drawn line is plotted, we observe that the maximum appears in the stain in contact with the mouth and the whole side in contact with the head (right) has an intensity equal or greater than the other side (Fig. 3.17).

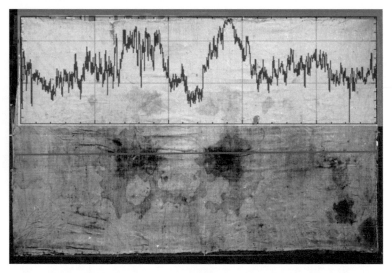

Figure 3.17 Solazzi transparency photograph of the Sudarium (Courtesy: Roberto Falcinelli from Archive Diocesan Center of Sindonology Giulio Ricci-Rom). Intensity of the stains along the line added in the upper part.

[12]Falcinelli, R. (2007). *Análisis sobre la tela de protección del Sudario de Oviedo. Cronohistoria de una investigación.* Proceedings of the 2nd International Congress for the Sudarium of Oviedo. pp. 635–650.

3.3 The Fabric

The Sudarium of Oviedo is a rectangle of 83 × 53 cm in average values. It does not present selvage in its edges, which indicates that the linen must have had a larger size previously. Already in the 1572, Ambrosio de Morales proposed that *"..., There is no doubt that the entire Sudarium of our Redeemer is not in this Relic, but only part of it, because it was not possible to wrap the entire Divine Head in this cloth alone"*. However, the size of the Sudarium is in fact enough to wrap the head if we put around it only a layer. To get it we must put the Sudarium horizontally i.e., along the standard arrangement and run the 83 cm around the head. It could be thought that Morales looked the Sudarium in "vertical" position and in this way the 53 cm are not enough to wrap the entire head.

The Sudarium has a texture of taffeta (or tabby). That is to say, a fabric in which the threads are woven in the simplest form: a thread of weft crosses above and below a thread of warp every time (Fig. 3.18). This type of fabric was in use since 5.000 BC and it is still used today.

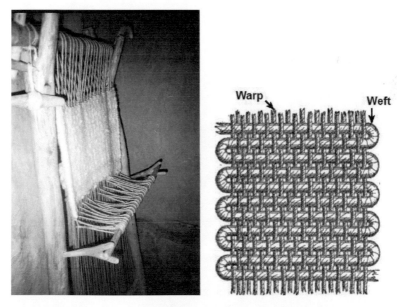

Figure 3.18 Roman loom and warp and weft in a taffeta fabric.

The total weight of the linen is 89.06 g, as it was measured in the sacristy of the Cathedral of Oviedo on 24 March 1998. Overall, this provides an approximate weight of 20 mg/cm^2.

The fabric of the Sudarium has an average density of 43 threads/cm in warp (40–45 threads/cm) and 19 threads/cm in weft (18–21 threads/cm), with a mean diameter of the threads of 0.24 mm (0.15–0.30 mm). Each thread is made of between 100 and 120 fibres. The fibres have diameters between 10 and 20 microns. The threads of weft are thinner and homogeneous and are as long as the largest length of the cloth. There is more variability in the diameter of the threads of the warp than in the threads of the weft. This is a common occurrence in manual spinning, and it is usual in ancient textiles.

The microscopic observation was made with a stereoscopic microscope with enlarging up to 60× and with SEM. The threads are linen (*Linum usitatissimum L.*), giving the colorations of Vetillard and Floroglucin positives. Figure 3.19 shows a transverse section of a linen thread of the Sudarium taken with SEM. No other natural fibres have been detected.

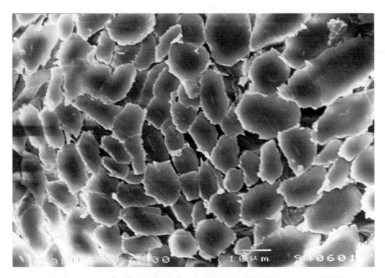

Figure 3.19 Section of a thread of the Sudarium (© F. Montero CES).

In general, the threads of any cloths made up by fibres can have a twist or spun S or Z (Fig. 3.20). The warp can have one spun and the weft, the other. Therefore, we can speak of system Z/Z, system S/S or system S/Z. In the case of the Sudarium, the threads are twisted in Z both in warp and in weft (Fig. 3.21). It is system Z/Z. The angles of twist are approximately of 24° in warp and 20° in weft. It is a weak spinning. It provides a soft fabric, not very resistant to abrasion, easy to wrinkle and to get dirty but it undergoes little shrinkage.

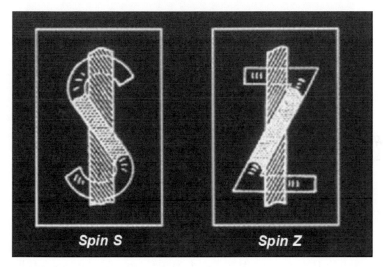

Figure 3.20 Forms of twisting the threads, S and Z.

Could these characteristics of the fabric of the Sudarium indicate a possible origin of its manufacture?

The spinning of the linen in S was used in Egypt from the first dynasties (5000 B.C.) until the 17th century[13]. The fibre of wet linen takes the S-spun spontaneously and, therefore, the S-spun is the most frequent observed. Linens fabrics found in Israel (including 70 in En-Gedi) were all of S-twist and plain weave, both simple type of one thread over another (taffeta or tabby) or double type or "basket weave" of two threads over two. Most of the wool fabrics were also S except for a small portion that presents Z twist (37 samples out of 826)[14].

[13]Vial, G. (1989). *Le linceul de Turín – Etude Technique*. CIETA – Bulletin 67. p. 13.

[14]Shamir, O. (2014). *A Burial Textile from the First Century CE in Jerusalem Compared to Roman Textiles in the Land of Israel*. Proceedings of the Workshop on Advances in the Turin Shroud Investigation. 4-5, September 2014 Bari, Italy.

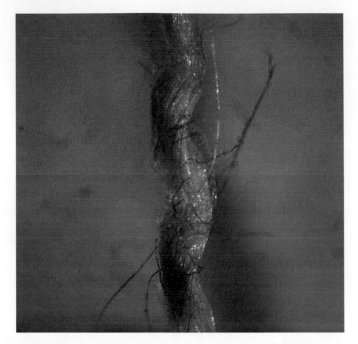

Figure 3.21 Thread of the Sudarium showing its spin in Z (© CES).

In the wide variety of fabrics found in the "Cave of the Letters" in Nahal Hever near the Dead Sea[15], out 33 fabrics of linen inventoried only one was twisted in system Z/Z (sample n° 69, 13 threads/cm and 11 threads/cm), whereas the rest were spun in system S/S. Other twenty linens explored later were S-spun[16]. In the Cave of the Rope (Nahal 'Arugot at the South of Jerusalem and the West of the Death Sea) other five linens dated to the Roman period showed the warp and weft threads in S-spun[17]. Wool textiles show S/Z or Z/Z spun in very few cases, but the rule is also S/S spun. This seems to confirm that in the Mediterranean oriental basin (Egypt and current Israel, Jordan, Syria) in the Roman epoch, the spun was S (system S/S) for the most part of

[15]Yadin, Y. (1963). The finds from the Bar-Kochba period in the Cave of Letters, Jerusalem.

[16]Shamir, O. (2018). *Textiles, threads and cordage from the cave of letters*. In Dead Sea. New Discoveries in the Cave of Letters. Series: Crosscurrents: New Studies on the Middle East. New York: Peter Lang. p. 191.

[17]Eitan, K., and Porat, R. (2016). *Nahal 'Arugot: Final Report.* Hadashot Arkheologiyot: Excavations and Surveys in Israel 128. p. 2.

linen as well as of wool and cotton. Only after the Medieval Age the spinning started to be in Z in the East.

The Z spun was typical however in the West (Greece and Italy) of the Roman Empire (27 B.C.–610 A.D.). The archaeological textiles of the Roman time discovered in the Mediterranean zone that show the use of system Z/Z presuppose a manufacture in Northern Europe and the Galia[18]. In the Iberian Peninsula, at that period, we find both types of twist. And it was frequent to twist the warp in S and the weft in Z (system S/Z), or vice versa to give a major cohesion to the fabric.

The Sudarium, which is made of Z-twist, was probably not made in Israel. This characteristic is shared with the Shroud of Turin. It is also made of linen twisted in Z, thus, indicating a manufacture abroad Israel. In fact, the last DNA analysis of the linen of the Shroud places its spinning in India[19]. The regular spinning in India was Z. In this country, cotton was used more frequently than the linen. Cotton has a natural tendency to twist in Z. It could be possible that the natives observed the cotton fibres themselves spinning in Z while drying and imitated the natural process. When the craft was established in India to twist in Z, it was used also for linen[20]. This reinforces the hypothesis that the Shroud was manufactured in India. Moreover, the presence of traces of cotton *Herbaceum* mixed with the linen threads indicates that they were spun using the same equipment on which cotton was spun,[21] corresponding to the ancient Indian practices. In the case of the Sudarium, Raymond Rogers also found a fibre of cotton among the linen fibres of the Sudarium[22].

---

[18]Cardon, D. (1999). *Archéologie des textiles : Méthodes. Acquis, Perspectives*, Minutes of Lattes's colloquium, October, 1999. Cited by Montero, Felipe. *Description of the Sudarium of Oviedo*. Proceeding of the first International Congress for the Shroud. Valencia 28–30 de April 2012. p. 8.

[19]Barcaccia, G., et al. (2015). Uncovering the sources of DNA found on the Turin Shroud. *Sci. Rep.* 5, 14184; p. 7.

[20]Batigne, R., and Bellinger, L. (1953). *The Significance and Technical Analysis of Ancient Textiles as Historical Documents*. Proceedings of the American Philosophical Society. Vol. 97, no. 6, p. 671.

[21]Raes, G. (1973). *Rapport d'Analyse*. Supplemento Rivista Diocesana Torinense, January. 1976.

[22]Roger, R. (2005). *The Sudarium of Oviedo: A Study of Fiber Structures*. Available in https://www.shroud.com/pdfs/rogers9.pdf (October 2020).

It should be reminded that the trade, including the imports of textile products, was flourishing during the Roman Empire. This has been demonstrated by the existence of large quantities of cotton fabric from India, system Z/Z, in the famous Berenike's port in Egypt where the usual spun system was S/S[23].

Additionally, the weave of the Shroud is a very elaborate 3/1 herringbone twill pattern. It is an expensive cloth. If it were the Shroud of Christ, it was bought by Nicodemus, a wealthy man and it must have been valuable according to the Jews customs for their deceased. Therefore, the fact that the Shroud was not manufactured in Israel is not a strong objection for its authenticity. It could have been imported by a buyer of a certain economic level. The Jewish shroud from the time of Christ found in Akeldama would have been imported as well[24]. If the Sudarium and the Shroud were not woven in Israel, but both cloths were used for the same corpse, they could both have been imported together by the same person.

An important fact that is often ignored is that the density of the fabric is considerably high. The weave of the Sudarium is very tight, being 43 × 19 threads/cm. The high density is another common characteristic with the Shroud of Turin. The number of threads per cm in the Turin Shroud is 38 × 27, which is very high compare to linen textiles manufactured at the Holy Land which usually have 10 × 15 threads per cm[25]. Other densities of Israeli linens in the Cave of Letters are 12 × 12 threads/cm[26]. The five linens in the Cave of the Rope show about 20 × 15 threads/cm in average. Only one of the five linens with 30 × 21 threads/cm[27] approaches close to the 43 × 19 threads/cm of the Sudarium.

---

[23]Wild, J. P., and Wild, F. (2014). *Through Roman eyes: cotton textiles from early historic India*. In: A Stitch in Time: Essays in Honour of Lise Bender Jørgensen Gothenburg University, 2014. pp. 209–236.

[24]Shamir, O. (2014). *A Burial Textile from the First Century CE in Jerusalem Compared to Roman Textiles in the Land of Israel*. Proceedings of the Workshop on Advances in the Turin Shroud Investigation. 4–5, September 2014 Bari, Italy.

[25]Shamir, O. (2014). *A Burial Textile from the First Century CE in Jerusalem Compared to Roman Textiles in the Land of Israel*. Proceedings of the Workshop on Advances in the Turin Shroud Investigation. 4–5, September 2014 Bari, Italy.

[26]Shamir, O. (2018). *Textiles, threads and cordage from the cave of letters*. In Dead Sea. New Discoveries in the Cave of Letters. Series: Crosscurrents: New Studies on the Middle East. New York: Peter Lang. p. 191.

[27]Eitan, K., and Porat. R. (2016). *Nahal 'Arugot: Final Report*. Hadashot Arkheologiyot: Excavations and Surveys in Israel 128. Table 1.

We have the weave density of several fragments of linen fabrics and copies of the Shroud of Turin. All the copies were elaborated with a painted figure between Chambery (France) and Turin (Italy). They are listed in the Table 3.1.

Table 3.1 Weave density for some fabrics

Origin	Year	threads/cm
Sudarium of Oviedo Spain	?	43 × 19
Shroud of Turin Italy	?	38 × 27
Shroud Fragment of Oviedo Cathedral Spain	?	25 × 24
Linen textiles from Land of Israel	1st Century	10 × 15
Shroud Fragment Toledo Cathedral Spain	<1250	33 × 26
Sudarium Fragment Pamplona Cathedral Spain	1255/58	21 × 21
Copy of the Shroud in Xabregas Portugal	≈1519	27 × 27
Copy of the Shroud in El Escorial Spain	1567	34 × 22
Copy of the Shroud in Valladolid Spain	≈1567	32 × 23
Copy of the Shroud in Navarrete Spain	1568	30 × 28
Copy of the Shroud in Plasencia Spain	>1578	18 × 18
Copy of the Shroud in Torres de la Alameda Spain	1620	23 × 23
Copy of the Shroud in Logroño Spain	1623	28 × 26
Copy of the Shroud in Arquata Italy	1653	27 × 26

Two observations can be made from these data. Weaves of the fabrics in the 16th and 17th centuries around the Alps have higher densities than in Israel in the Roman Period. The Sudarium and Shroud have more threads in one square centimetre than any other of this list. The weave density of threads issue should be studied.

The weave of the Sudarium of Oviedo presents some manufacture defects. There are loops of yarn formed in threads of weft, probably due to an attempt of recovering the horizontal position of the weft. There are also crossings of parallel and contiguous weft threads, possibly due to the existence of two weavers because the loom was quite broad, and that the weft thread was not launched by a single shuttle but by two shuttles from each edge (Fig. 3.22).

Figure 3.22 Crossing of weft threads for two weavers (© CES).

All crossings of the threads of weft practically happened at a distance of 31–33 cm from one edge and 52–50 cm from the other edge. If the crossing happened at the middle of the fabric, the original width of the cloth should have been 100–104 cm. Then, a part of about 19 cm would be absent (Fig. 3.23). Although this width can be about two royal Talmudic cubits, the lack of selvage in the right edge (standard arrangement), indicates that it was not the edge in the loom. Then, under the assumption that the crossing happened at the middle of the fabric, the Sudarium should have been wider in the loom. We will see later that the cloth could have also been wider at the time that was used for the deceased.

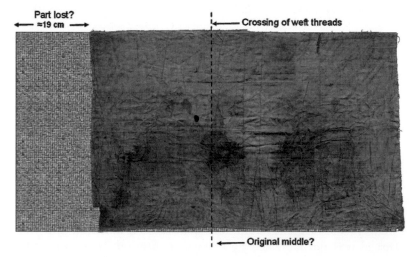

Figure 3.23 Possible Sudarium part lost.

The types of defects and the other characteristics of the fabric point to an old cloth that has been made by hand, in a vertical loom with weight. The Sudarium could be part of another larger fabric that was cut after it was taken from the loom. The type of loom does not determine the date or the place of manufacturing because vertical loom with weight was in use since 4.000 years ago and continues to be utilized today.

3.4   Other Traces

In the Sudarium, there are more than 180 perforations of different size (0.2–0.8 mm) that go through the fabric. These have been caused by a sharp object like a needle that has separated the threads leaving a conical hole. Usually, the holes appear forming couples, one with the conic form left by the sharp object when it entered and the other with the conic form left by the sharp object when it exited (Fig. 3.24).

Figure 3.24 Couple of conic holes in and out (© CES).

Furthermore, it can be seen in many of them how the lateral threads are displaced (Fig. 3.25), as if they had supported a tension. This characteristic is also present in the holes where the thread that sew the Sudarium in modern times to the backup cloth passed through. The displacement of the threads to a side of the holes indicates that this tension had to be pulling enough time to leave this permanent deformation of the threads or the tension was strong enough although of short duration. Some of the holes coincide with areas stained with blood.

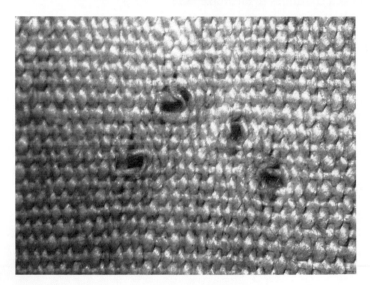

Figure 3.25 Holes with sign of tension (© CES).

This kind of holes is found in the areas close to where there was hair on the head or face (Fig. 3.26). It supports the conclusion that the Sudarium was sewn to the hair of the corpse. The holes circled in blue are surrounding some folds in the fabric and they caused these folds. The holes circled in red correspond to stitches on the hair. The stitching of the Sudarium to the nape hair was the first and involved a single layer. The stitching of the Sudarium to the beard involved the two layers of the cloth. To help understand these explanations, Fig. 3.27 shows a coarse scheme of where were the face and the nape in contact with the cloth (left half of the figure) and where the "soaked" face was passed through to the second layer of the folded Sudarium (right half of the figure). A detailed justification of this outline will be found in following Section 4.4.

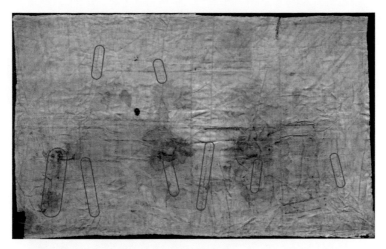

Figure 3.26 Places of holes on the hair areas.

In addition, a group of the holes run in pair, very close to one another, along both sides of a fold, being the cause of it. This type of holes is marked in blue in the Fig. 3.26. One of them in the lower part of the cloth corresponds to the stitching to the beard. We can see an enlargement in the Fig. 3.28. This group passes through a blood stain. The other two groups of holes marked in blue in the upper part of the cloth correspond to stitches to the hair.

Figure 3.27 Scheme of face and nape on the Sudarium.

Figure 3.28 Parallel Lines of holes along the sides of a fold running on the area where the beard was.

If we look at the lower left corner of Fig. 3.26, which corresponds to the nape area, we can see other holes surrounded in red also in parallel lines but more separated and less grouped in pairs. They are shown in detail in Fig. 3.29 with illumination by transparency. This set of holes was used to sew the lock of hair of the nape. The confirmation that the holes were caused by the stitch with threads came when a remaining thread was found in the upper occipital area still in its place (Fig. 3.30). It joins several of these coupled holes. This piece of cord was made of three linen threads twisted in "Z" and twisted themselves together in "S" to give them cohesion (Fig. 3.31).

Other small end of thread was found in November 2006 near the edge opposite to the place of lock hair (Fig. 3.32). At this place there was not probably hair or beard to sewn them. This thread or cord is in turn made of two threads also of linen but twisted each thread in S and then twisted both in S as well. Either two types of threads were used maybe by two different people to sewn the Sudarium to the head[28] or this last thread is a

[28]Montero, F. (2012). *Description of the Sudarium of Oviedo*. Proceeding of the 1st International Congress for the Shroud. Valencia 28–30 de April 2012.

contamination from modern stitches. This second possibility comes because it was inside a hole and the place of the strand of thread where it was found is close to the edge that was sewn several times to the support cloth near to the wooden frame in the past.

Figure 3.29 Parallel Lines of holes on the area where the nape was (© CES).

The places of stitches in Fig. 3.26 lead us to the following reconstruction of the use of the Sudarium. It was sewn to the hair and beard of the corpse. First, it was sewn to the lock of the hair behind the nape around a thick quantity of hair, and then, it was deployed along the left side of the head over the left ear to reach the face up to the right cheek. There, it was fixed with stitches to the beard and to the hair near the forehead. These last stitches are closer in pairs, causing a remaining fold in the fabric. The rest of the Sudarium was folded back to cover the face with a second layer. This second layer was sewn too to the beard and to the hair near the left ear.

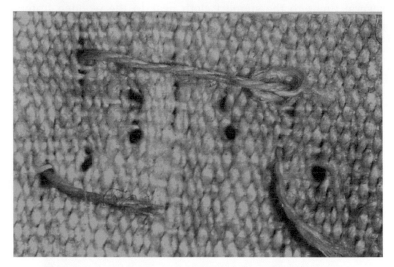

Figure 3.30 The thread on the upper occipital area (© CES).

Figure 3.31 Twist of the thread (cord) in the occipital area (© CES).

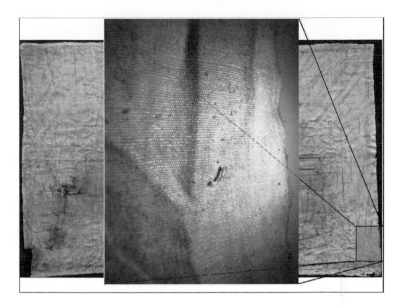

Figure 3.32 Thread on the right lower corner.

Because the two layers in front of the face are stained with a mixture of blood and oedema, its objective could have been to soak these body fluids to avoid them getting lost on the ground. In the Jewish culture, the blood of the deceased must be buried with the body. The Jewish friends of the cadaver, fearing that the blood would fall to the ground, may have covered up the face with two layers to prevent the loss of that blood.

3.5 Small Blood Scabs

On the entire surface of the Sudarium, mainly on the side that was in contact with the corpse head, there is a multitude of blood micro-scabs located mostly on the upper part of it (Fig. 3.33). It can be seen that some of them are reaching the very edge of the canvas (Fig. 3.34), which confirms that the original cloth was larger than the current one at the moment of its use on the corpse. This indicates that the cloth has been in contact with small bleeding wounds existing on the top of the head.

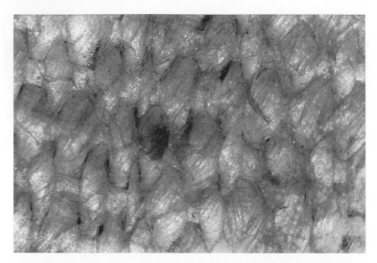

Figure 3.33 Blood scab corresponding to the upper part of the head (© CES).

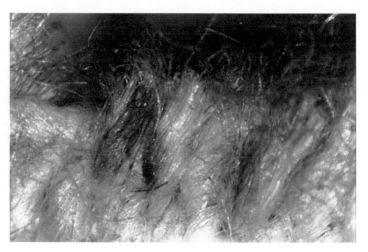

Figure 3.34 Blood scab at the cloth edge (© CES).

The observation through optical microscopy on both sides of the Sudarium either of the main blood stains as in the corner stain, the so-called butterfly stain, and also in the stains of wounds produced by sharp objects revealed that the entry of blood into all of them come from the dirtiest side of the fabric (Fig. 3.35).

Figure 3.35 Both sides of the blood stain at the lower corner (dirtiest side at the right).

This is the conclusion because the existence of a higher blood concentration in the threads on this side can be observed and also because of the presence of blood corpuscles in the crossing of these threads. This indicates that the liquid blood fraction has passed by capillarity action through the fibres of the threads from one side to the other, yet the coagulated part has remained on the surface as if the fabric were a filter that has retained them. This tells us that this side is the one that was in contact with the emitting sources responsible for the formation of these stains.

3.6 Folds

The Sudarium is dirty and wrinkled over its entire surface (Fig. 3.36). It is well known how easy it is for linen fabrics to be wrinkled and how difficult it is to remove these wrinkles later. However, all of them provide valuable information about the causes or origins of their formation. These folds can be classified into two types or groups:

(a) The folds that are formed in the fabric by folding it into eight parts to store the fabric.

(b) Those formed by placing the cloth on the bleeding head, trying to adapt to it, and which have been formed at the same time as the stains.

Regarding the first group a) and looking at the direction of the folds and whether they protrude or sink with respect to the surface of the cloth, we can say what their formation order was when the Sudarium was folded. If we look at the side that was in contact with the head, the sudarium was folded in half vertically two times; one backward and then one forward symmetrically. One of these vertical fold creases matches the tear in the top of the cloth. Once the four vertical surfaces of 21 × 53 cm have been formed, they were then folded horizontally by the middle, forming a rectangle of approximately 21 × 27 cm (Fig. 3.37). In 1957, the Dean already realized that in previous times, the Sudarium was folded because it had some wrinkles horizontally and others vertically[29].

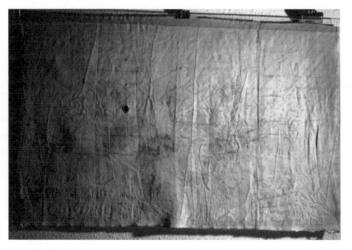

Figure 3.36 The great quantity of wrinkles in the Sudarium (© CES).

Regarding the wrinkles due to the adaptation and holding of the Sudarium on the head b), the first of greater magnitude is the one near the right of the first vertical fold of folding (Fig. 3.37 up). It runs also in vertical direction throughout its height. It is the axis of symmetry of the two main blood stains.

---

[29]Cuesta, J. Dean. (1957). *Guía de la Catedral de Oviedo*, Oviedo 1957, p. 126.

In its lower part it forms an inverted V with a vertex between and at the level of the darkest stains.

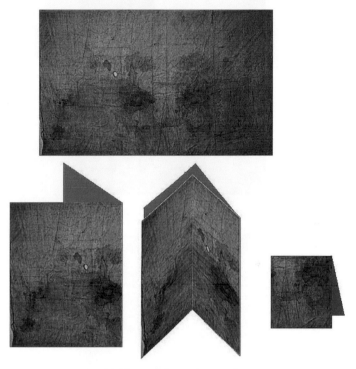

Figure 3.37 Sequence of folding of the Sudarium for saving.

Now we remember the group of aligned holes already shown in Fig. 3.28. The fold being protruding at the side that was in contact with the face indicates that the cloth was bent by this fold backward; leading the two main stains to coincide on the opposite side (Fig. 3.38).

Another wrinkle that is worth mentioning is the one at the lower left corner when looking at the standard arrangement. It is in the area where the back of the neck was placed. The fold passes by the middle of a blood stain called the "butterfly" stain because of its shape but it does not reach the "pointed stains" in the occipital area. It protrudes to the side of contact with the head (Fig. 3.39). The stain is almost symmetrical around the crease, but the stitches are enough apart. The shape of the stain

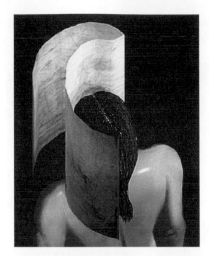

Figure 3.38 Folding of the Sudarium adapting to the head (© CES).

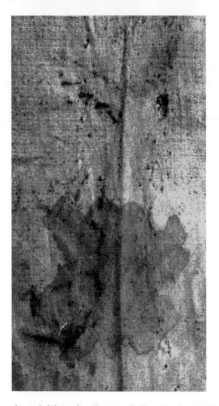

Figure 3.39 Protruding fold in the "butterfly" stain (© CES).

corresponds to a lock of hair stained with the mixture of blood and edema and wrapped with this part of the Sudarium sewn with the two stitches lines around the lock of hair. However, it is not easy to reconstruct the arrangement of this part of the Sudarium when it was fixed on the head of the corpse. When we simulate the direction, the fold protrudes (Fig. 3.40), we do not find evidence on where the wrapped bundle hair was because at least a part of the cloth should be between the head and the hair.

Figure 3.40 Direction of the fold on the nape.

There is other group of creases in the upper right corner when looking at the standard arrangement of the Sudarium. There, we can see there a series of wrinkles almost parallel and in diagonal direction that correspond to those left by a possible knot in that corner (Fig. 3.41).

When we extrapolate the lines of the wrinkles to converge to a single vertex, it appears obvious that there is a lack of fabric. The size of this lack can be about 20 cm on each side (Fig. 3.42). As we said above (Fig. 3.23), if the fabric was larger by the right edge (standard arrangement) as the knot suggests, the hypothesis that the crossings of the threads of weft happened in the middle of the full width in the loom implies now a new size between 138 and 142 cm which does not correspond to any royal Talmudic cubit. In fact, the result of a simulation of knotting a cloth similar to the Sudarium at the top of the head shows nearly parallel wrinkles like the wrinkles of the Sudarium (Fig. 3.43). However, it was not larger than the Sudarium. In conclusion, it is not at all necessary to have a larger cloth to mark wrinkles as seen in the Sudarium when we tie a knot.

Figure 3.41 Folding created by a knot (© CES).

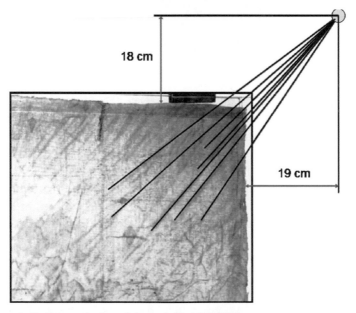

Figure 3.42 Extrapolation of the wrinkles (© CES).

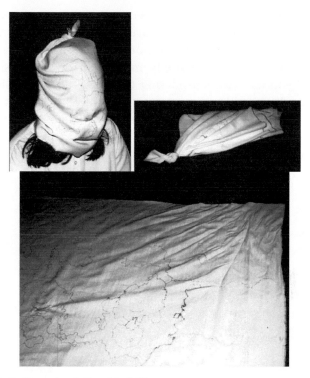

Figure 3.43 Knot simulation (© CES).

This knot matches the paraphrase of Saint John given by Nonnus of Panopolis as we saw in Section 2.3. The singularity of this knot referred by Nonnus and its presence in the Sudarium of Oviedo, supports a dating before the 5th century.

3.7 Contamination

Felipe Montero[30] and the other members of the EDICES found a series of external contaminations not inherent to the makeup of the fabric. The observation was made on the entire surface of the cloth with magnifications between 10 and 60, taking macro-photographs. There are colonies and fungal spores, pollen, the presence of wax drops, burns, carmine traces, red ink, remains of silver paint, remains of adhesives, environmental dust, tears, etc.

---

[30]Montero, Felipe. (2012). *Description of the Sudarium of Oviedo*. Proceeding of the 1st International Congress for the Shroud. Valencia 28–30 de Abril 2012.

The dust fills the gaps between the threads, being more abundant in the blood-stained areas (Fig. 3.44). It is striking that the degree of contamination on the side that was in contact with the corpse is considerably more than the other side and especially in its folds. Considering that the relic has traditionally been displayed without protecting the visible surface, it can be thought that the public has revered it for quite some time for the side that was in contact with the corpse (standard arrangement)[31].

Figure 3.44 Dust between the threads (© CES).

In Section 1.1, we introduced the silver[32] paint contamination. It corresponds to the bottom of the can (170 × 80 mm) that was placed on the side that was in contact with the head since the threads are crushed on this side under the weight of the can and there are practically no traces on the other side.

The presence of silver does not happen only in the stain of the can. Some micro particles scattered on the cloth and mainly on the top of the folds (Fig. 3.45) show the attribute of silver when analyzed under SEM using an Energy Dispersive

---

[31]Montero, Felipe. (1994). *Descripción química y microscópica.* Proceeding of the 1st International Congress for the Sudario de Oviedo. Oviedo. 29–31 October 1994. pp. 67–82.

[32]The presence of silver was determined by EDS (Energy Dispersive X Rays) analysis.

X-ray microprobe. The makeup of these micro particles includes Silver, Sulfur, Aluminum, Carbon, Oxygen and traces of Calcium. A possible explanation for the origin of this surface contamination with Silver is a long-time rubbing of the cloth with a silver reliquary. If the silver contamination is homogenous over the entire surface (not reported), the hypothetical reliquary should be large enough and expensive! If the silver contamination is only important over a partial quadrant, the hypothetical reliquary could have a size like DIN A4, or even much smaller depending on the way of folding. But there is no documented reference to such a box. The reliquary hypothesis faces the objection that the expected way to store a relic in the antiquity is to protect it inside other fabric of silk or of other nature but avoiding the direct contact between the reliquary boards and the relic. Could these silvered particles be related to the event when the paint can was placed on the Sudarium?

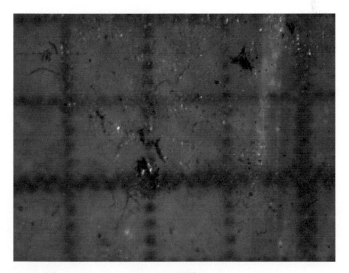

Figure 3.45 Silver micro particles (© CES).

Wax drops can be also seen on the dirty side, as well as the perimeter of a hole which was due to a burn (Fig. 3.46). The analyses carried out showed that it was from beeswax. It cannot be very ancient since those that were used in Roman Empire were made with animal tallow and the beeswax candles were introduced later.

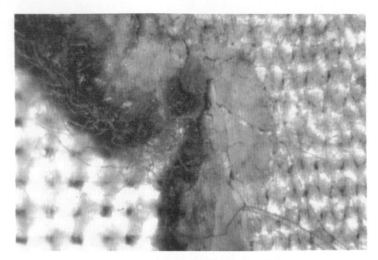

Figure 3.46 Wax in the perimeter of a burned hole (© CES).

Also, on the same dirty side, there are red stains on the surface of the threads that are due to carmine remains logically received during some exhibition of the Relic to the worshipers (Fig. 3.47).

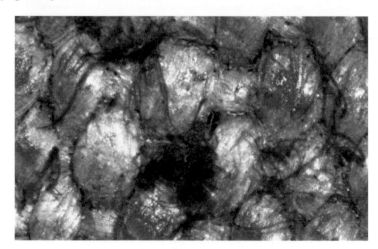

Figure 3.47 Carmine on the threads of the Sudarium (© CES).

To collect the dust adhered to the cloth; the EDICES team used an adjustable vacuum pump, coupled to a Millipore system with a Teflon membrane of 5 microns pore size and 47 mm of diameter (Fig. 3.48). The vacuuming was made with a very small depression in order not to damage the cloth and without touching its surface. Each time, the vacuumed surface was of around 5 cm × 30 cm.

Figure 3.48 Collecting particles with a vacuum pomp (© CES).

This procedure is much cleaner and harmless to the cloth than the procedure used occasionally for the Shroud of Turin. In addition, this method provides with samples not only of the surface but also of the deeper parts of the fabric. Ten membranes were used on the whole in two missions to cover almost the whole surface of the Sudarium. The three first membranes were protected with adhesive tape. However, for the other seven, the assembly was improved using hermetic plastic sample holders that allow the direct metallization of the samples, as well as their subsequent direct microscopic observation on the surface of the membrane. The adhesive embedded the particles making morphological observation difficult, essential for the classification of pollen grains.

With this method, a significant number of particles were captured, both inorganic (sand, silica, micro meteorites, etc.) and organic (carbon, pollen grains, fungal spores, fibres, insect remains, etc.).

Microspheres of 8 microns in diameter were also found, which, when analyzed by energy dispersive X-rays, revealed that they were made of iron. Other particles were identified as clay.

Among the organic particles, some red blood cells in very small numbers were discovered too.

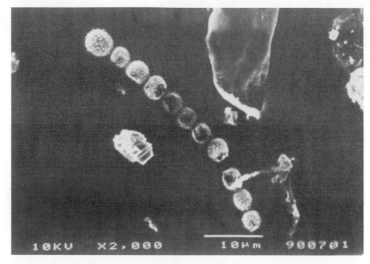

Figure 3.49 Conidia of spores (© CES).

Figure 3.50 White resins particles of aloe and balsams of storax on the blood-stained threads (© CES).

Some of the fungi spores were loose and others forming chains (conidia. Fig. 3.49), showing in some cases their mycelia although they were generally much damaged. Other abundant group of particles (5 to 15 microns) of organic origin that are not pollen grains (Fig. 3.50) deserve special attention because of its significance. They are described in the following subsection of Aloe and Storax. A particular description of the study of pollen is also included in its own subsection later in the chapter.

3.8   Aloe and Storax

The particles that were nicknamed as potatoes by Felipe Montero, led to think that they could be remains of some type of resin of aloe and myrrh just when they showed their shapes under the SEM electronic beam (Fig. 3.51). They were analysed by the members of the EDICES Rafael Sánchez Crespo and Josefa Prada Álvarez-Buylla. The sample was put in a 1: 1 benzene/methanol mixture, subsequently methylated, using the Meth-Prep II reagent for the gas chromatography/mass spectroscopy analysis. A large number of peaks could correspond to terpenes from aloe resin.

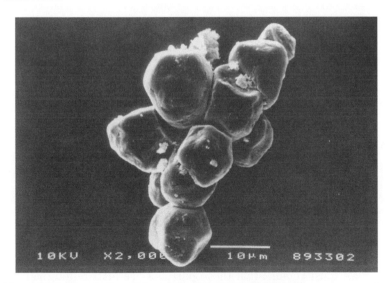

Figure 3.51 Group of particles of aloe ("potatoes") (© CES).

In the monosaccharide analysis the sample was compared with a standard of aloe and myrrh. The monosaccharide distribution was like that of myrrh and aloe, although the amount of mannose detected in the sample was very low compared to the one of the two standards.

The presence of many contaminants in the sample did not allow a clear identification of sugars by direct observation of the chromatogram. Despite this, the selective search for the ion m/e = 115 and 139 allowed obtaining a distribution of sugars very similar in qualitative composition to that of myrrh. However, the presence of myrrh could not be confirmed. The presence of aloe was also possible, even though the sugar content was lower.

After identifying more than two hundred compounds, the results show that these particles are made of: Diterpenoids–Manoyl oxide–Cinnamic acid esters.

The studies were carried out by these researchers over three years. They clearly indicated that these compounds may have their origin in aloe resins and storax balsams (*Liquidambar orientale*). The latter was a compound that replaced myrrh because it was cheaper. This product was largely produced in Palestine, and it was exported to Egypt[33].

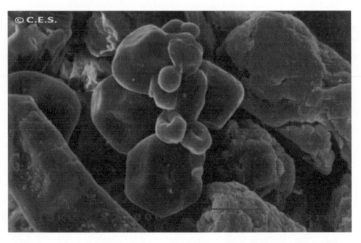

Figure 3.52 Red blood cells on the aloes and storax particles (© CES).

We must remark that these particles are distributed over the entire surface of the cloth. Besides, they are especially concentrated in the areas where there is more blood. Additionally, they

[33]Montero, Felipe. (2007). *Descripción química y microscópica del Lienzo*. Proceeding of the 2nd International Congress for the Sudarium of Oviedo. Oviedo 13–15 April 2007. p. 123.

appear on top of the blood, which allows us to deduce that the substance was scattered across the cloth when it had already been used, since, otherwise, they would remain under the blood. Figure 3.52 displays one of the best photos showing red blood cells on those particles that have served to preserve them through the centuries.

3.9   Pollen

The analysis of the pollen grains has been carried by several researchers over the years. Before the arriving of the EDICES, Max Frei took samples in 1979 under suggestion of Mr. Ricci using segments of adhesive tape and he provided the following report (Table 3.2).

Professor Max Frei reached partial conclusions like some of the studies carried out later by members of the EDICES (see below). He found sixteen types of pollen in the Sudarium of Oviedo and identified at least nine of them growing in Palestine[34]. This was a point of agreement with the Shroud of Turin where he found more types as indicated in Table 3.2.

Frei reported on this occasion 55 species of plants in the Shroud of Turin. In other reports, he counts up to 58[35]. To locate the Sudarium in Holy Land, we should pay attention to the four species in the upper list of the report (Table 3.2). The rest of the plants grow in the Mediterranean area as well as in other regions and they only provide evidence that the Sudarium was in Spain which is something already known. Among the four species in the upper list, three of them also have presence in the Mediterranean area according to the Global Biodiversity Information Facility (GBIF)[36]. However, the species of *Acacia albida* is not recorded in the entire Mediterranean area except in

---

[34]Ricci, G. (1985). *L'Uomo della Sindone è Gesù*, Ed. Cammino, Milano. p. 234–238.

[35]Marion, A., et al. (1997). *Nouvelles découverts sur le Suaire de Turin*, Albin Michel. This list does not include four of the sixteen detailed in the Sudarium report: *Phoenix dactylifera, Ceratonia siliqua, Parietaria diffusa* and *Plantago coronopus* which are Mediterranean species, and they are not interesting to study the geographical origin of the Sudarium.

[36]https://www.gbif.org/. This is an international organization whose data are provided by many institutions from around the world. Their data are primarily plants distribution and other biological species in the world.

Israel (Fig. 3.53). It was only one pollen species among the sixteen reported. If we trust the Frei study, it provides evidence that the Sudarium, as well as the Shroud, were in Israel before they arrived in Spain and France, respectively.

Table 3.2 Max Frei report of the pollen found in the Sudarium of Oviedo and in the Shroud of Turin [see Note (a)]

GROUP OF PLANTS WHICH ARE FREQUENTLY FOUND IN PALESTINE IN THE DESERTS, SALTY PLACES, OR ON ROCKS		
	Shroud	**Sudarium**
Acacia albida Del.	+	+
Hyoscyamus aureus L.	+	+
Phoenix dactylifera L.	o	+
Tamarix africana Zoh.	o	+
	23	4
GROUP OF MEDITERRANEAN PLANTS WHICH GROW ALSO IN PALESTINE		
	Shroud	**Sudarium**
Ceratonia siliqua L.	o	+
Cistus creticus L.	+	+
Phyllirea angustifolia L.	+	+
Pistacia lentiscus L.	+	+
Ridolfia segetum moris	+	+
	16	5
GROUP OF LARGE AREA PLANTS IN EUROPE THAT GROW IN ITALY, FRANCE, AND ASTURIAS (SPAIN)		
	Shroud	**Sudarium**
Corylus avellana L.	+	+
Parietaria diffusa Mert. à K.	o	+
Plantago coronopus L.	o	+
Taxus baccata L.	+	+
	12	4
Plants not identified so far	4	3

(a) "+" type of pollen present in the cloth and "o" type of pollen not present in the cloth among the total indicated below.

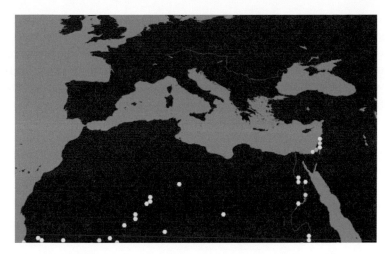

Figure 3.53 Geographical distribution of *Acacia albida* Delile (yellow spots). Ref. GBIF.

In the Shroud's case, this pollen cannot have reached this cloth if it was manufactured in France and travelled to Italy. In short, the presence of this pollen could clearly indicate that both cloths, Shroud of Turin, and Sudarium of Oviedo, may have had a common origin in Israel.

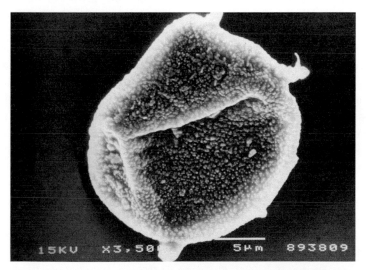

Figure 3.54 A pollen grain found on the Sudarium (X 3500) (© CES).

Using the SEM and a light microscopy, a more extensive study was carried out by Carmen Gómez Ferreras[37] on behalf of the EDICES. She analysed 141 grains deposited on the filter inserted in the vacuum device (Fig. 3.48). Some of them were well preserved with characteristic shapes and ornaments (Fig. 3.54).

The researcher of the EDICES found 25 types of pollen. Her information differs from the Frei's report because she did not observe the *Acacia albida* and she only found the *Tamarix africana*, among the four more decisive species reported previously. This last species also grows in the Mediterranean area; therefore, it is not an evidence of the origin of the Sudarium in Israel. Only five out of the sixteen species found by Frei were confirmed by Gómez Ferreras:

> *Tamarix africana*
> *Pistacia lentiscus (terebinthus)*
> *Corylus avellana*
> *Parietaria diffusa (judaica)*
> *Plantago coronopus*

We highlight the difference between genus and species that Gómez Farreras uses. The genus includes several species. Different species of the same genus can grow in different regions. All the 25 species identified by the Spanish EDICES member belong to Mediterranean categories. On the other hand, Gómez Farreras proposes possible geographic indicators of the presence of the Sudarium in Palestine. However, to use the pollen as geographical indicator, it is not enough to identify the genus. In order to do this, it is necessary to know the endemic species. Although they were not found among the grains she analyzed, inside the genera *Pistacia* and *Tamarix*, there are endemic or native Palestinian species. They are *Pistacia palestina* and some species of *Tamarix*. The rest of the genera of pollen found in the Sudarium are widely distributed across and outside the Mediterranean region. In the 1st Congress of the Sudarium of Oviedo (1994), the species identification by comparison with the corresponding control preparation was pending. It is unknown whether this last task was ever performed.

[37]Gómez Ferreras, C. (1994). *El Sudario de Oviedo y la Palinología*. Proceeding of the 1st International Congress for the Sudario de Oviedo. Oviedo 29–31 October 1994. pp. 83–90.

In the 2nd International Congress of the Sudarium of Oviedo (2007), another presentation about pollen took place. However, Maria José Iriarte, only explained the theoretical research without any further data.

A more recent study on the pollen subject was published in 2016 by Marzia Boi[38] and presented in the Pasco Conference in 2017. She undertook the study analysing with SEM the vacuumed powder. Boi took a new approach to the pollen subject. She does not focus on its role as a geographic indicator but in its significance according to the embalming and funeral rituals about 2000 years ago. She proposes that some of the traces found on the Sudarium as well as in the Shroud of Turin have their origin in the burial substances used for the entombment. In fact, *Helichrysum* pollen would have been present in oils and ointments because *Helichrysum* oil is only produced from the pressing of fresh flowers and it is not an airborne pollen (carried out by the air).

Relevantly, the Director of the EDICES, Alfonso Sánchez Hermosilla during the analysis of blood stains, discovered an entomophilous pollen of the type of *Asteraceae*, just identified as *Helichrysum*. It was found attached and embedded in dust[39]. This discovery confirms that pollen became adhered when the fluid was still fresh and not yet solidified, demonstrating an original presence from the very first moment that the blood contacted the fibres of the burial cloth. The Sudarium pollen has been compared with the referential *Helichrysum* (Fig. 3.55[40]) and apparently, there is a great similarity of the types. It seems impossible that pollen could offer so much information, yet it is the only element that with certainty can reach the present day without being completely destroyed, and it is the only component that can be identified exactly as it was.

---

[38]Boi, M. (2017). Pollen on the Shroud of Turin: The Probable Trace Left by Anointing and Embalming. Archaeometry. 59(2). pp. 316–330.

[39]Catholic University of Murcia (UCAM) https://international.ucam.edu/university-news/ucams-researchers-have-found-scientific-evidence-places-shroud-oviedo-and-shroud-turin-same. (2015).

[40]Grain of the Sudarium (left) from UCAM. Control samples (right) from Boi, M. (2017) *Pollen on the Shroud of Turin: The Probable Trace Left by Anointing and Embalming.* Archaeometry. 59(2), 316–330.

Figure 3.55 *Helichrysum* pollen grains from the Sudarium (left) and a control sample (right) under scanning electron microscope.

The presence of the *Helichrysum* on the Sudarium is therefore an evidence of its funeral use. According to Boi, this pollen may also be misunderstood as *Gundelia Tournefortii* on the Shroud of Turin.

3.10 Dust

This section recovers the main work presented in the workshop in Bari 2014[41]. In March 2012, Rodrigo Álvarez, member of the EDICES at that time, performed X-ray fluorescence measurements on the Sudarium of Oviedo inside the Cathedral. The primary goal of the X-ray fluorescence detection is to estimate the relative amount of some chemical elements through the different areas of the cloth. During the test, a characteristic photon is emitted, allowing identifying the source element of the Periodic Table. A portable Fluorescence X-ray Fluorescence analyzer Niton XL3t was used. This equipment is endowed with a 45 mm^2 Silicon Drift Detector. The voltage selected was 50 kV with a current of 40 μA. Measurement time for each spot was 60 seconds[42]. The analysis was performed on 57 spots of the side of the

---

[41]Barta, C., et al. (2014). *New Coincidence between Shroud of Turin and Sudarium of Oviedo*. Workshop on Advances in the Turin Shroud Investigation (ATSI). September 4, 2014.

[42]For a complete description of the instrumental and experimental procedure see Barta, C., et al. (2014). *New Coincidence between Shroud of Turin and Sudarium of Oviedo* already referenced.

Sudarium that was in contact with the head (standard arrangement) whose distribution was predetermined. These points included blood stains and clean areas. Figure 3.56 shows the location of every measured spot.

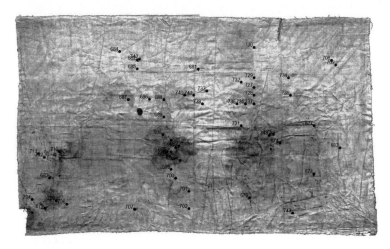

Figure 3.56 Location of points measured on the Sudarium of Oviedo with XRF of Oviedo.

The aim of the XRF application was to find elements that could be constituents of the mineral particles of the dust and of the various liquids present in the fabric. Elements with atomic numbers greater than 16 (S, sulphur) can be detected by the equipment in the actual experimental conditions. In this case, a good reliability for the measurement of calcium (Ca) and potassium (K) was found. The obtained data are mainly useful for relative comparison among the quantities detected in the different spots.

It was interesting to separate the measurement on the blood stains from those on the "clean areas", mainly in the periphery. The differences between the concentrations detected in the clean and stained areas are statistically significant only for Strontium (Sr) and Ca. Additionally, the higher concentrations of K and S detected on the stained areas confirm the physiological nature of the stains. The most evident difference between the stained and the clean areas is the content of Ca, which was always higher in the former. The excess of the particles with Ca on the stained areas were probably fixed to the cloth by the physiological fluids while they were still fresh. The relatively high Ca content in the stains is significant, since both the stains and, therefore, the Ca are linked to some anatomical parts of the deceased man.

The author participated in this research where we highlight an important finding. We identified a particular spot with the highest quantity of Ca. We do not consider it a mistaken lecture as the two closest spots show the second and the fourth highest content of Ca (see Fig. 3.57).

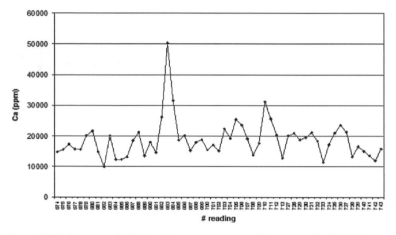

Figure 3.57 Content of calcium for every measured spot on the Sudarium.

The three mentioned spots belong to the stained group and are located close to the tip of the nose. On the other hand, calcium can be also associated to soil dirt. This supports the assumption that the body was lain face down for some time. The nose is an atypical part of the body to present this singular dirt. Analogously, as we will see in a further chapter, just, an unexpected excess of dirt around the tip of the nose was also detected in the Shroud of Turin.

We will now present the search of the possible origin of the limestone retained in the Sudarium. To this purpose, we considered the presence of Ca. Calcium is obviously quite common in limestone soils, yet two hypotheses deserve consideration: the limestone can come both from the Calvary and Oviedo's Cathedral (Fig. 3.58).

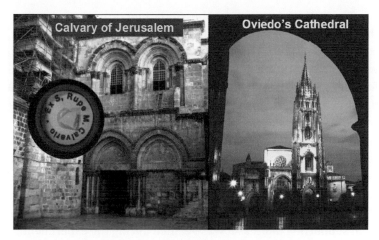

Figure 3.58 Two possible origin of the limestone of the Sudarium.

Several samples of the Calvary rock and the stones of the Oviedo's cathedral were analyzed using the same portable detector. Concerning the X-ray fluorescence analysis, Calvary limestone appears to have very few impurities.

Apart from Ca, the Sr detected in our analysis can be used as an indicator of the type of limestone deposited on the Sudarium since it does not have a biological but mineral origin. The ratio Sr/Ca in the Calvary limestone is particularly low compared to limestone from different geographical places (Site of the Baptism in Jordanian, Spanish, and Caribbean beaches, etc.), including that from Oviedo's Cathedral. We can assess the similarity between the ratio Sr/Ca detected in the Sudarium and between the Calvary and Oviedo's Cathedral. Table 3.3 shows the different Sr/Ca ratios calculated for the Calvary limestone, Oviedo's Cathedral limestone and for the Sudarium (separating between stained and cleaned areas).

Table 3.3 Ratio of Strontium to Calcium for Sudarium and Limestone from Oviedo Cathedral and Calvary

Origin	$Sr/Ca \times 10^3$
Cathedral limestone	2.43
Calvary limestone	0.24
Clean area of Sudarium	0.75
Stained area of Sudarium	0.63

The ratios detected in the Sudarium are more similar to the Calvary's ratio than to the Cathedral's ratio, suggesting a stronger similarity with the former. As it was explained before, we consider that the fluids, when still fresh, acted as cement for the environmental dust. Therefore, the stained area should have a higher proportion of mineral particles from the place where it was used. The ratio in the stained area of the Sudarium is even closer to the Calvary limestone ratio. These results lead to the hypothesis that over 75% of the limestone deposited in the Sudarium comes from the Calvary attributing only the rest to the Cathedral. In addition, the level of Sr detected in the cloth of Oviedo can be overestimated as some of it may come from supporting table, as it was verified during the experiment.

At the time of the workshop in Bari (2014), it was found that the iron (Fe) distribution detected did not correlate with the intensity of the blood stains. This fact deserved further research. In the following months we realized that the support in which the measurements on the Sudarium were performed had a decisive impact on the result. That is, in absence of any support behind a dummy cloth when the measurement is performed, the presence of Iron is well correlated with the amount of blood staining the dummy cloth (Fig. 3.59). The impact of the support raises also doubts about the reliability correctness of the comparison between rate Sr/Ca in the Sudarium and in the stones of the Calvary and Cathedral, since the measurement setup was different for the stones of the Cathedral and the Sudarium. As a new test was never performed, the reliable results are only those linked to the different distribution of dust (Ca) along the areas of the Sudarium.

Additionally, a secondary confirmation deals with the stains of the painting container contamination. Three points were measured on the stains of the paint container. The XRF analysis did not include silver detection, but the signal for mercury (Hg) at these three points—and only for these three—was significant confirming the painting-related nature of these stains.

As a conclusion from the X-ray fluorescence analysis, we can highlight that the calcium shows a statistically-significant higher presence in the areas with bloody stains. It was probably fixed to the cloth when the physiological fluids were still fresh or soon after. This fact allows correlating its distribution with the anatomical features of the corpse. The highest content of Ca is observed close to the tip of the nose. However, to assess the possible origin of the dust present on the Sudarium, additional test should be performed.

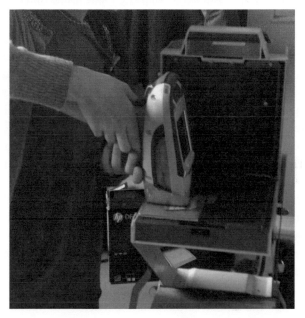

Figure 3.59 Test of XRF on a dummy cloth to verify the impact of the support.

3.11 Summary of the Scientific Analysis

Between the second half of the 20th century and the second decade of the 21st century, the arrangement of the Sudarium of Oviedo underwent several changes. It was fixed in a silver decorated frame, showing a side or the opposite after two interventions: Before the documented changes, the dirtiest side was visible. The forensic study concluded that this side was the one in contact with the head of the corpse. It was the visible side when Mons Ricci "discovered" the Sudarium in 1977. It was turned around to show the cleaner side at some instance prior 1987; probably taking place during an observation of the Ricci team in 1986. In 1987, nuns removed nails and repaired the assembly. The cleaner side remained exposed to the visitors until 2006, when the EDICES suggested that the important side to be exposed was the dirtiest. This landscape format is referred to as the **standard arrangement** for reference.

In April 2014, the conservation conditions were improved by placing the Sudarium inside an urn sealed with Nitrogen atmosphere.

The infrared, ultraviolet and transparency photographs allowed discriminating the nature and intensity of the blood stains. It revealed the presence of contamination as wax and modern lint fibrils.

The textile analysis identifies linen as the fabric nature and points out to a manufacturing outside Israel, probably in India. The Sudarium preserved in Oviedo was part of a larger piece manufactured in a loom.

Accurate observation revealed the stitches along the areas where the man had hair, supporting the theory that the Sudarium was sewn to the hair of the corpse.

The folds show that there were two layers in front of the face of the man and the evidence of an ancient knot in a corner of the Sudarium over the upper part of the head.

Among the most interesting particles of contamination recovered from the surface, we remark the red blood cells, the Aloe and Storax (a substitute of Myrrh) and the pollen.

The *Acacia albida* Delile grain found by Max Frei provides evidence that the Sudarium was in Israel before arriving in Spain. Additionally, the *Helichrysum* grain embedded in blood found by Sánchez Hermosilla is an evidence of the funeral use of the Sudarium according to Marzia Boi.

The analysis of the X-ray fluorescence shows higher presence of dust in the areas with bloody stains, locating the highest value close to the tip of the nose.

Chapter 4

Blood on the Sudarium

The first question to be evaluated in this chapter is the nature of the purported blood stains. In case of a positive result, the question would be whether it is human blood. Later, we will search for the specific characteristics of the blood. Finally, we will try to reconstruct the mechanism of formation and we will describe every type of "blood" stain.

4.1 Identifying Blood

The presence of human blood on the Sudarium of Oviedo has been verified by different specialists throughout the years. The first analysis was carried out by Baima Bollone[1] who in 1985 directly took seven thread ends from the clean areas of the Sudarium as well as thread ends from bloody spots[2]. He performed a generic blood diagnosis by microscopic observation and by three identification approaches:

---

[1]Professor of Legal Medicine at the University of Turin.

[2]Bollone, B., and Pastore, F. (1987). Proceeding of IV National Congress of Studies on the Shroud. Syracuse. 17–18 October 1987. Ed. Paoline. Torino 1988. Bollone, B., et al. (1994). *Resultado de la valoración de las observaciones y de los exámenes de algunas pruebas efectuadas sobre el Sudario de Oviedo*. Proceeding of the 1st International Congress of the Sudarium of Oviedo.

The Sudarium of Oviedo: Signs of Jesus Christ's Death
César Barta
Copyright © 2022 Jenny Stanford Publishing Pte. Ltd.
ISBN 978-981-4968-13-3 (Hardcover), 978-1-003-27752-1 (eBook)
www.jennystanford.com

– Detection of inorganic constituents of blood by means of Rx micro-spectrometer coupled with scanning electronic microcopy
– Thin-layer chromatography using the Farago A technique
– Incubating bundles of fibres in accordance with the Teichmann procedure

This last method is based on the decomposition of haemoglobin when it is combined with sodium chloride and soaked in acetic acid solution to crystallize in haematin hydrochloride. Despite the simplicity of the method, it has the maximum reliability for positive detection[3].

Regarding the seven samples, four were located on bloody spots and the other three on clean spots (Fig. 4.1).

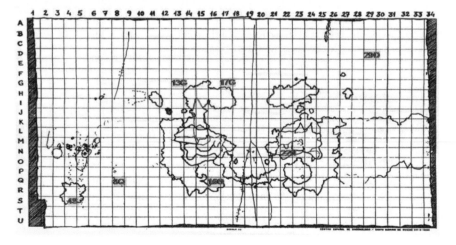

Figure 4.1 Location of the seven samples used by Bollone(© CES).

The three samples from clean spots gave negative results as expected (29D, 13G and 17G). Three samples taken from the bloody spots gave positive result (8Q, 22N and 16O). The 4S from the "butterfly" stain gave negative result against what was expected.

---

[3]Castillo, N., and Martinez, S. (2020). Teichmann Confirmative Test for Blood Identification in Stains. Scientia et Technica Año XXV, Vol. 25, No. 01, marzo de 2020. Universidad Tecnológica de Pereira. Colombia. p. 159.

The list of results is as follows.

29D	– (negative)
13G	– (negative)
17G	– (negative)
4S	– (negative)
8Q	+ (positive)
22N	+ (positive)
16O	+ (positive)

In addition to the tests, the microscope allows us to locate the human red cells[4]. They show their characteristic donuts shape (Fig. 4.2). The microscope shows red cells, *the most abundant type in the blood. They have no nucleus, and this fact immediately means that they are not from fish, amphibians, reptiles, or birds, since all those do have a nucleus. It is also possible to see their round shape, meaning that they do not come from mammals related to camelids (camels, dromedaries, llamas, alpacas, guanacos), whose red cells have an elliptical shape[5].* Therefore, for Bollone, the correlation of positive/negative test results is enough to confirm the microscope observation.

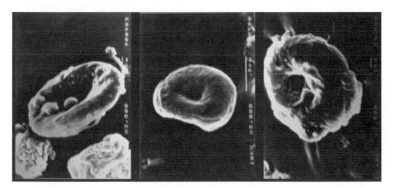

Figure 4.2 Three red cells found in the Sudarium of Oviedo (© CES).

The Bollone team also progressed by determining if the species of the blood corresponded to the human. The positive and negative results for these last tests matched the previous generic

[4]Bollone, B. (2009). *El Misterio de la Sábana Santa.* Algaida. p. 139.
[5]Sánchez, Hermosilla. A. (2012). *The Oviedo Sudarium and the Turin Shroud.* I International Congress on the Shroud. Valencia. p. 8.

tests. The conclusion of the Bollone team was that the Sudarium is a linen cloth that contains stains of human blood and pollution with other materials.

The next specialist who undertook the blood analysis was Dr Carlo Goldoni[6]. First, he presented his results in 1993 in Rome[7] and a year later, in Oviedo[8]. In 1992, Msgr. Giulio Ricci gave Goldoni a fragment from the Sudarium of Oviedo (Fig. 4.3) with the specific task of finding out whether there was any presence of human blood. The fragment came from the so-called corner of Ricci, near the butterfly stain. He also had available to him some fragments of the supporting canvas which was in close contact with the Sudarium for more than 400 years. The complete fabric of support was given to Ricci as a gift by the Council of Canons on one of his visits to Oviedo and was later deposited in Rome.

He conducted the research on three fragments:

- A sample of support linen (negative control).
- A fragment of the Sudarium of Oviedo of 28.2 mg.
- A sample of support linen soaked in fresh blood (positive control).

On the three types of fabric samples, without separating the possible blood from the linen, Goldoni applied the following tests:

- Screening with Ames test
- Test with Benzidine
- Reaction with Adler-Mayer reagent (for haemoglobin detection).
- Identification under microscope of acid haematin Teichmann crystals
- Identification under microscope of haematoporphyrin crystals (reddish fluorescence under blue violet illumination)

Goldoni included in his set of tests, some different ones from those carried out by Bollone, increasing the reliability of the identification.

---

[6]Haematology Specialist in Rome.

[7]Goldoni, C. (1993). *Human Blood on the Sudarium of Oviedo?* II International Scientific Symposium on the Turin Shroud.

[8]Goldoni, C. (1994). *Estudio Hematológico sobre las muestras de sangre del Sudario tomadas en 1978.* 1st International Congress on the Sudarium of Oviedo. Universidad de Oviedo. Oviedo.

Figure 4.3 Fragment of Sudarium tested by Dr Goldoni (© CES).

In the case of the Sudarium of Oviedo, he verified the positive reaction and although weak, it was evidently different from the clean control. The reactions for the control samples were as expected: intense for the sample soaked in blood and fully negative for the clean control sample. These tests confirmed the presence of haemoglobin (Fig. 4.4).

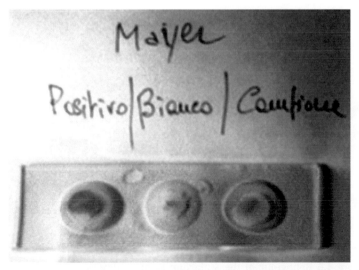

Figure 4.4 Haemoglobin positive reaction of the Sudarium sample (right) (© CES).

As it happens with Bollone, Goldoni completed the test with species detection. He used horse immunoserum to discover the immuno-globulins IgG that fluoresce under ultraviolet light. The result was also confirmed the presence of human globulin bound to the flax fibres (Fig. 4.5).

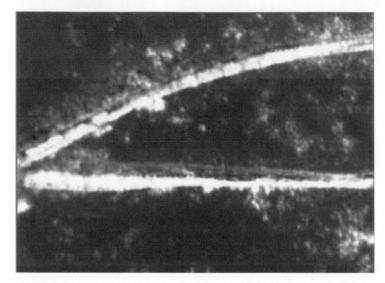

Figure 4.5 Human globulins IgG bound to the flax fibres of the Sudarium (© CES).

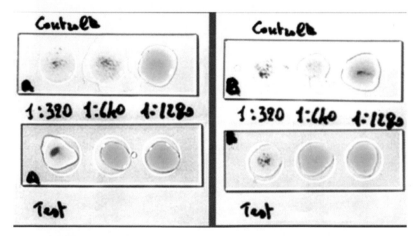

Figure 4.6 Behaviour of the Sudarium sample (down) against anti-A and anti-B agglutinins (© CES).

Goldoni undertook the identification of the blood group ABO knowledgeable about the difficulties that this task would involve. He found the presence of the typical agglutinogens A and B. Moreover, he was not able to observe anti-A or anti-B agglutinins adhered to the linen fibre (Fig. 4.6). Therefore, it behaves in the same manner as linen soaked with AB group blood like in the Holy Shroud in Turin.

He concluded by suggesting that AB could be the blood group of the Sudarium (being present in the corpuscular part of the agglutinogens A and B, and absent in its serous part of the anti-A and anti-B binders). However, he acknowledges that bacterial or other pollutions may disguise the presence of group agglutinogens inducing false reactions (mainly the "antigen B"), while the same immunoglobulins found in the linen might have kept their antigen power without retaining the power to react with specific antigens.

He limited his last conclusion to affirm that the Sudarium presents human blood on its surface. Nothing less!

The following study of the blood on the Sudarium was performed by Dr Villalaín[9] a member of the EDICES. He used crusts of blood from the pointed stains of the nape, a stained fragment of the fabric from the "Ricci corner" close to the piece used by Goldoni and some threads and marginal fragments of the cloth[10]. He replicated the generic diagnosis obtaining a positive reaction to the benzidine. He applied the additional method of Lecha-Marzo and Piga that dissolves the sample in acetic acid and ammonia. Six slides were prepared for tests with the dissolution:

Slide 1-Teichman reaction in the Bertran variant.
Slide 2-Stryzowsky reaction.
Slide 3-Reaction of Sardá.
Slide 4-Takayama's reaction.
Slide 5-Lecha-Marzo reaction.
Slide 6-Guarino reaction.

---

[9]Professor of Forensic Medicine. University of Valencia. Vice president of CES.
[10]Villalain, J. D. (1994). *Naturaleza y formación de las manchas*. 1st International Congress on the Sudarium of Oviedo. Universidad de Oviedo. Oviedo. pp. 139–140. A summary in Heras, G., et al. (1993). *El Santo Sudario de Oviedo.* Proceeding of the II National Congress of Paleopathology (Valencia, October 1993).

All of them gave a positive reaction and the spectral analysis corresponded with blood components, confirming again, the presence of blood.

The next step was the species diagnosis. Several traditional methods failed because the stains were insoluble due to their age. Then, the Villalain team used the same method as Bollone. They applied anti-total human immunoglobulin and anti-human IgG, on stained fibres and clean fibres. In both cases, the sample was positive in the stained areas and negative in the controls as expected.

After this third confirmation, it can be finally stated that the stains are of human origin.

Villalain with other specialists[11] searched for the determination of the ABO group and the tests were applied on stained samples and clean samples as control. Stained samples indicated weak group A and positive B. However, the clean samples also gave positivity reaction to group B. To verify if the group B was a false positive, they carried out a quantitative assessment of agglutinogens in the stained and clean areas, using the modified Holzer technique. The differences were significantly in favour of the stained fibres; consequently, the B antigen must also be admitted as a characteristic of the blood stain. The doubts expressed by Goldoni about the reaction to the B group get weaker because of this last test.

The blood group, therefore, may be considered AB according to the possibilities offered by the forensic haematology. Nevertheless, the serological data can only go so far.

The blood group is a test to identify individuals. But, to get to its objective we need to know the blood group of the person being investigated. In this case, this individual is Jesus Christ and we do not have a direct sample of his blood. We can only compare with the blood of other relics attributed to him.

4.2 Identifying DNA

DNA analysis is progressively substituting the determining of

---

[11]Dr Ruiz de la Cuesta from the laboratory of the Forensic School in Madrid.

blood groups as a method of individual identification.

In the congress of 2007, Dr Antonio Alonso[12] presented the analysis of DNA present on a stained thread and on a clean thread of the Sudarium[13]. In 1994, members of the Institute of Forensic Sciences took three threads from two of the most stained areas of the Sudarium (central stains), as well as the threads from adjoining non-stained areas (to be used as white controls). It means a total of 12 samples. The samples were kept frozen at –80°C in the institute. A first unsuccessful attempt was performed in 1994 with a stained thread and a clean thread. The other ten threads remained frozen in the Institute. It was only in 2005, when the identification techniques had improved, that two other threads were used (one stained and another clean) for new analysis. Figure 4.7 shows the location where the threads were extracted from. The objective was to assess the presence of human nuclear and mitochondrial DNA as well as the presence of human haemoglobin in the presumed blood remains.

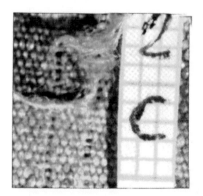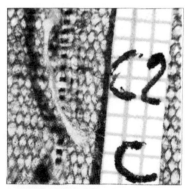

Figure 4.7 Location of the extracted threads for the DNA analysis (© CES).

The detection of human haemoglobin by immune-chromatography test was negative. If we considered the previous three positive results for detection of human haemoglobin, we must acknowledge that the method used by Dr Alonso was not sensitive enough. The nuclear DNA quantification was also

[12]Director of the National Institute of Toxicology and Forensic Sciences.
[13]Alonso, A., et al. (2007). *El ADN del Sudario de Oviedo*. Proceeding of the 2nd National Congress on the Sudarium of Oviedo. pp. 167–173.

negative.

For the quantification of mitochondrial DNA, the comparison between the stained sample and control sample was significantly positive with 50 couplets/ml for the stained sample as opposed to 10 couplets/ml for the control sample. Through the PCR technique, they managed to obtain mitochondrial DNA fragments of the HVI region for the stained sample and they edited a complete sequence of the HVI region. On the contrary, they failed to obtain positive results for the clean sample used as blank (Fig. 4.8). This means that in the blood there is human mitochondrial DNA. However, the specialist cannot exclude the possibility that the DNA comes from some form of modern contamination. Dr Alonso did not disclose the actual DNA sequence to avoid further biased tries to obtain the same sequence and to allow future blind tests.

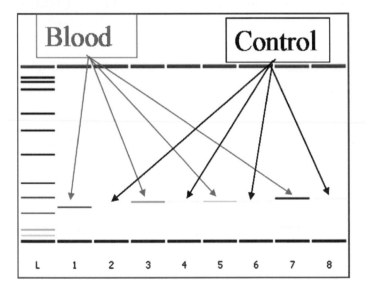

Figure 4.8 Mitochondrial DNA fragments of the HVI region presents in the blood sample and absent in the control sample.

While doubts cannot be ruled out, we have the complete DNA sequence of the HVI region for the blood of the Sudarium. Although the significance of this sequence was not recorded in the proceedings of the 2007 Congress, during the talk, Dr Alonso said that the sequence found has an important identification value because it only appears in one case out of more than 8000 when checked against European databases. Now, the question is if it is possible to obtain the sequence of the same HVI region from other samples related to Jesus Christ.

Dr Alonso also found DNA of the fungi *Arachnomyces minimus* that grows in rotten wood. Other analyses show that there is no living mycological activity.

4.3   Presence of Fibrin

Fibrin is a fibrous, non-globular protein involved in the clotting of blood. In the case of internal or external bleeding, a series of signals trigger a cascade reaction with the final result being the generation of fibrin implied in the mechanism of the formation of blood clots. If the fibrin acts on damaged tissues poor in blood vessels, such as the pleural, pericardial, or peritoneal membranes, fibrin clots are formed despite the scarce presence or even the absence of blood.

On 9 March 2012, Dr Alfonso Sánchez Hermosilla[14] made a direct examination of the Sudarium of Oviedo with a reflecting microscope at 500X. Among many other observations that confirmed data already established, he detected the presence of structures that are compatible with dried blobs of fibrin (Fig. 4.9).

They did not appear to be associated with blood cells under UV light showed no fluorescence. That leads us to some important conclusions[15]. The size of the blobs of fibrin found cannot be formed within the circulatory system or in bleeding wounds or in the blood of a corpse before putrefaction. The most likely hypothesis is that such blobs were formed within a body cavity, pleural and/or pericardial. To reach such a condition, the individual must undergo a severe trauma as the scourging and a few hours must elapse to allow the fibrin formation before the fluid is released. The fibrin accumulated in the pleural cavity

---

[14]Director of EDICES since 2013. Forensic. Legal Medicine Institute of Murcia - Cartagena Branch.

[15]Sánchez Hermosilla, A. (2012), *The Oviedo Sudarium and the Turin Shroud*. 1st International Congress on the Shroud. Valencia. p. 37.

needs a path to reach the Sudarium. This path could be the wound from the spear that connected the cavity with the respiratory tracts.

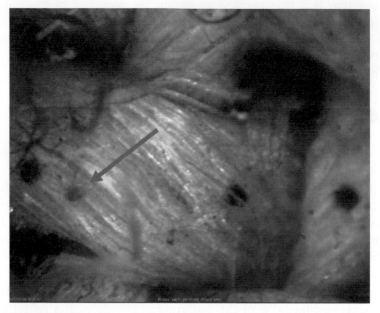

Figure 4.9 Microscopic blob of fibrin in a thread of the Sudarium of Oviedo (© CES).

In the Sudarium of Oviedo, the presence of these fibrin blobs free from blood elements could be another indirect proof of the scourging, because the formation of fibrin requires a previous injury. This is the most plausible hypothesis for the formation of the fibrin elements.

To reach the Sudarium, the fibrin could flow out through the nose and the mouth, together with blood and the liquid of the edema from the lungs.

4.4 Mechanism of Formation of the Bloodstains

The sequence of the formation of the stains on the Sudarium was the most original and decisive contribution of the late Dr José Delfín Villalaín, already referenced previously as a

Professor of Forensic Medicine at the University of Valencia and Vice President of the Centro Español de Sindonología. The analyses were realized, first in the Laboratory of Criminology and Forensic Biology at the School of Legal Medicine in Madrid and, later, in the Faculty of Medicine in Valencia[16].

Before going any further, we have to define some names for the stains according to their morphology and characteristics (Fig. 4.10):

- Main stains that are symmetrical. They include the background stains and the more intense central stains.
- Pointed form or dot-shaped stains near the left edge.
- Butterfly form stain near the lower left corner.
- Stain of the lowest left corner.
- Diffuse stain between the pointed form stains and the main stains.
- Accordion stain on the right side.

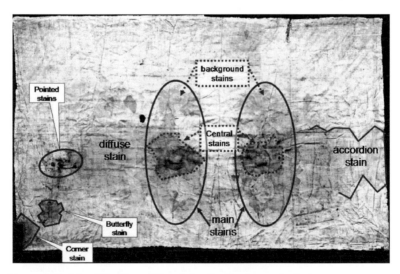

Figure 4.10 Identification of the stains for the dynamic study.

4.4.1 Main Stains

Regarding the mains stains, there are the background stains with

---

[16]Villalain, J. D. (1994). *Naturaleza y formación de las manchas.* 1st International Congress on the Sudarium of Oviedo. Universidad de Oviedo. Oviedo. pp. 131–176.

a diluted bloody appearance that surround the central stains which are darker. Their aspect cannot be attributed to the passing of time, even though blood stains tend to lighten, transforming the red colour of the blood into brown; the tone of the stains suggests that they are diluted blood to a greater or lesser extent.

The next step was to determine the degree of dilution and its nature. To this goal, experimental stains were carried out on white linen cloths, of a consistency like that of the Sudarium, with samples of whole blood, washed red blood cells in suspension in physiological saline, haemolysed red blood cells, serum, albumin, and blood at different dilutions (1/2, 1/4, 1/8, 1/16, 1/32), deposited on a stretched and horizontal cloth, with the amount being 5 cm^3 of each product in a series of ten (see an example in Fig. 4.11). They had to analyse 2420 stains to complete all combinations of the different factors. Once dried, at room temperature, they were compared with the photographs of the Sudarium.

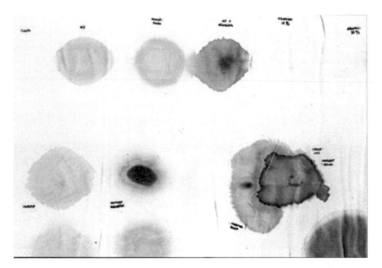

Figure 4.11 Stains of serum, albumin, and blood at different dilutions (© CES).

From the preliminary phase, it was found that the dilution was between 1/4 and 1/8. New comparative series established that the blood dilution was around 1/6. With the same experimental series, it was deduced that the background stains of the Sudarium of Oviedo are produced by serum and blood diluted to the proportion of approximately 1/6.

The central background stains behind the main stains appear on both sides of the axis of symmetry and are optically denser and with marked edges. To simulate its characteristics, a new series of whole, fractionated and diluted blood stains was used. However, in this case inert particles (carbon), and live yeast particles were added in variable quantities. Again, the systematic series were deposited on linen cloth, like that of the Sudarium. It was found that the morphology of these spots was produced by blood diluted to the same concentration (1/6), on which there were various cell clusters, which sprouted in successive waves (Fig. 4.12). It was observed that this mixture simulated fingerprints well, the origin of which could be some fingers trying to contain the flow from the mouth and nose.

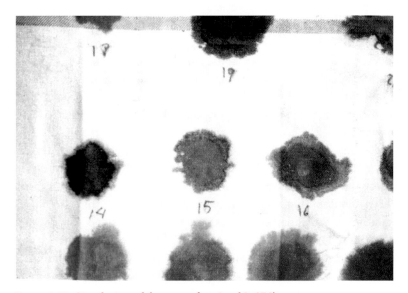

Figure 4.12 Simulation of the central stains (© CES).

The darker and denser central stains are structured "by layers" in a clearly manifested succession. Considering its morphology and delimitation, the specialist estimated that there have been between four and seven impregnations, which overlap in successive waves with a common central origin.

Almost all of the remaining stains were of the same nature. The diffuse stain and the accordion stain cover the entire width of the fabric with well-defined upper edges but have a more discoloured tonality.

The stain with the shape of butterfly wings and the stain in the missing corner have a morphology like the first symmetrical stains.

On the contrary, the dot shaped stains near the left edge (the wounds of the crown of thorns) differ from any other. Their morphology in the form of concentric halos around a denser nucleus deserves specific analysis.

Finally, there are multitudes of isolated and small stains.

The question now is to estimate the time elapsed between the formation of the superimposed main stains considering their morphological characteristics and to determine the order of their formation. In our case the background stains are clearly different from the central stains implying that a certain time has elapsed between one and the other, sufficient for the total or partial drying of the stain that previously originated.

To specify this point, a new series of successive experimental stains was carried out with different concentrations for the stains and variable elapsed time periods between their superposition. For this, six series of 30 blood stains were made, forming rectangles of 5 by 6 stains, depositing 0.5 cm^3 of liquid diluted to 1, 1/2, 1/4, 1/8, 1/16 and 1/32. On each of them, stains of these same concentrations were deposited, at 0, 15, 30, 60 and 120 minutes (Fig. 4.13). It was again verified that the blood dilution of the Sudarium stains was between 1/4 and 1/8 and that the time elapsed between the formation of the first background stain and the central stain was between 45 and 60 minutes. The tests were performed at the temperature and humidity of the laboratory in Madrid and were repeated, with similar results in the Valencia laboratory. The successful results can be seen in Fig. 4.14 and Fig. 4.15 where the obtained stains superposed after 45 minutes and 60 minutes, respectively, are compared with some contours of the stains in the Sudarium.

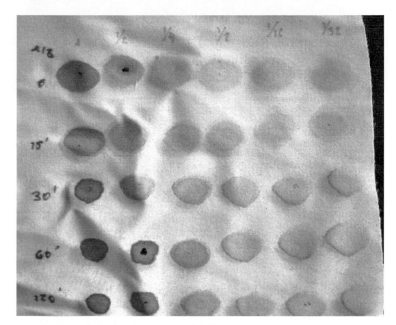

Figure 4.13 Simulation of superposition of stains (© CES).

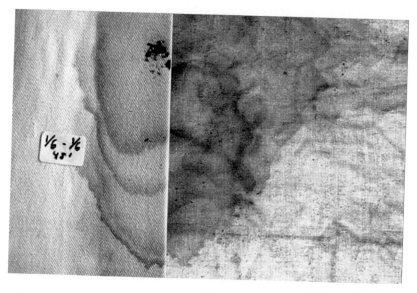

Figure 4.14 Comparison of superposition of stains after 45 min (© CES).

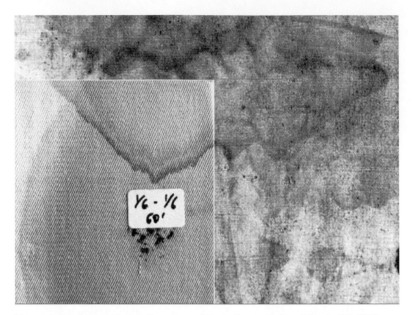

Figure 4.15 Comparison of superposition of stains after 60 min (© CES).

In terms of their formation, we can clearly differentiate between the background stain and the central stain which is in fact a complex of stains. The first stain is homogeneous, and it is produced by a mechanism of soft and progressive impregnation. On the contrary, inside the central stains, the overlapping conglomerate is not separated with precision. Their contours are less defined, and a certain pressure and multiple obstacles intervene. We have to assume a complex dynamic, repeated in different and more frequent waves, and that the source of the corresponding blood was at the top. Guillermo Heras[17] found the fingerprints from the fingers of a left hand which tried to stem the bleeding.

The background stains are larger and clearer and appear on both sides of the axis of symmetry. The stain increases its definition from the inside (left) to the outside (right) which indicates that the liquid was filtering in that direction, retaining the cell clusters through double filtering. The symmetrical conformation shows that the Sudarium was folded back on itself.

---

[17]Engineer. Director of the EDICES until 2013.

If we look at Fig. 4.10, the mixture of blood enters by the left part of the axis of symmetry and exits to the right of it.

These stains conform to the facial features of a corpse. To progress with the reconstruction, the tradition of how the cloth was used on Jesus Christ, when he was crucified, needs to be considered. This offers two possibilities: that the cloth had been placed on the head of the crucified, still on the cross, or once he had descended from it. In the first case, the head of the crucified was tilted, strongly flexed on the chin. In the second case, the corpse should be lying down. To test these hypotheses, Dr Villalaín proceeded with the head of a mannequin, to which he adapted hair, a beard and mustache and a nasal drip system, tilting it in all possible directions (Fig. 4.16).

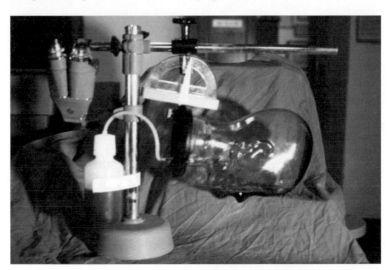

Figure 4.16 Mannequin to test the stains formation (© CES).

For the following analysis, we refer to the stain at the left of the axis of symmetry. In the standard configuration of Fig. 4.10, the left and right of this stain correspond also to the same anatomical sides. The specialist had to reconstruct the lower part of the stain that corresponds to the mouth, beard and chin. A trustworthy explanation is that the chest of the corpse was in inspiration (breathing-in pose) and with abundant pulmonary edema, hanging, with the head tilted downwards. But he also had to explain the higher part of the stain which corresponds to along the nose and to the forehead. After numerous attempts, they found one possible explanation for the formation of this stain: a double mechanism in which, while the body is hanging, the lower part of the stain is produced (Fig. 4.17) and then, in a continuous way, the middle and upper part of the stain is produced when the corpse descends and lies in the prone position (Fig. 4.18).

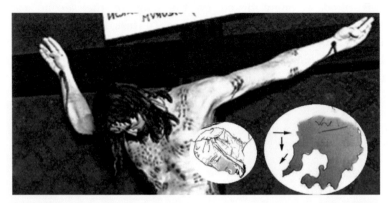

Figure 4.17 Formation of the low part of the main stain during the first position.

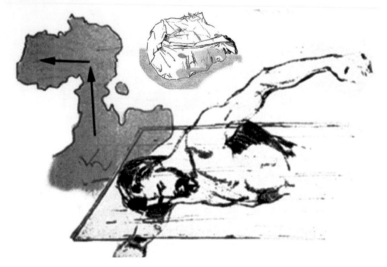

Figure 4.18 Formation of the high part of the main stain during the second position.

As postmortem relaxation occurs and as the early and progressive stiffness sets in, the pulmonary edema begins to flow smoothly through the mouth and nose: the cheekbone and tip of the nose remained at the same level; the edema soaked the mustache, beard and due to the inclination of the head, the right cheek and the cheekbone. The head of Jesus was bent to the right according to the asymmetrical distribution of the blood stains corresponding to the mouth with a predominant diffusion of liquid towards the right. This can be interpreted because the diffuse outflow of blood from nose and mouth followed the force of gravity. Consequently, the vertical axe of the head had to be tilted to the right. This does not necessarily mean that the head was tilted to the right with respect to the body. The shoulders, neck and trunk could be tilted to the right keeping the head aligned with the rest of the body. Most of the fluid flowed through the nose because it was verified that the flow through the mouth does not soak the mustache nor the beard as it appears in the Sudarium. A first conclusion was that the victim whose head was enveloped by the Sudarium was already dead.

However, the stain corresponding to the higher part of the nose, and to between the eyebrows and to the right forehead could not be produced in the previous position because that would go against the laws of gravity. It was necessary to increase the tilt the aforementioned mannequin forwards. Then, automatically, the nasal drip slipped on the back of the nose and flowed "upwards", reproducing the bridge of the nose and the forehead. However, the formation of the forehead stain falls to the right of the face, that is, to the opposite side of the stain over the left cheekbone. To obtain the two parts of the stain simultaneously, the head had to be flexed downward and slightly turned to the right in the cross. When turned to the right, the left cheekbone remained at lower level than the right cheekbone and the fluid soaked more of the left side. Some minutes later, the head had to be flexed downward and turned to the left on the floor. This way the right side of the forehead got more soaked than the left. This explains the asymmetry of the spillage of the left side of the beard and moustache, being more extensive than the right one, and that of the forehead that covers the right half but not the left. Only in this manner, when the mannequin system is tilted forward, is there a moment when the left cheekbone and the right forehead are at the same level and the nasal flow is distributed between the two. The elapsed time in the first vertical position was estimated to be about 45–60 minutes as determined above. The corpse must have remained in the second horizontal position for another 45–60 minutes.

According to this reconstruction, the line-up of the arms along the sides of the thorax, when the descent occurred would be not possible. The reason is because had it been so, the flow of the pulmonary edema would have been violent and gushing and that is not the characteristic of the background stains we are studying. On the contrary, an intense flow is more characteristic of the central stains. Therefore, in the hypothesis presented, the hanging corpse had to be taken down, whilst maintaining the original posture, that is, without the mobilization of the arms. It would support the hypothesis that the corpse was taken down, jointly with the horizontal beam, without detaching his hands from it until a later moment. The stain corresponds to the force of gravity added to the diffusion forces of the liquid that slowly penetrates the tissue.

Later, according to the aforementioned chronology, subsequent mobilizations produced several waves of fluid in the central stain that we have mentioned. This is quite a common experience in the postmortem phase of people who die with a "lung flooding" since, sometimes spontaneously and others when dressing or handling the corpse, pulmonary fluid comes out through the natural orifices. The central stain is formed with a certain pressure that tries to contain the flow. This might have originated at the time of the transfer and manipulation of the corpse for its shrouding and burial.

The researchers have proposed an explanation for the folding of the Sudarium that provides a double layer in front of the face. The early reason was the assumption that the right arm prevented the cloth from passing by the right ear because it was pushed against the shoulder. However, this is not the position when the head is tilted down with the arms fixed above the head. It is easy to verify in every sculpture of a crucified, for example, Fig. 4.17, that both sides of the head could be accessed. The simplest reason for folding with a double layer would be to contain the flow of blood. It is evident that a single layer did not seem to be enough because even the second layer was soaked. The goal of recovering the blood from a corpse to avoid its loss is typical of the Jewish culture, in which the blood of the deceased must be buried with the corpse. Thus, it was necessary to retain the blood and deposit it in the sepulchre.

In the final phase, the Sudarium must be opened to surround the whole head. This statement can be made, based on the study of the stains that are not on the four sides (reverse and obverse of the two halves), and, therefore, could not be produced with the cloth folded on itself. The best hypothesis is that the corpse had to be moved face up for a short path of 30 or 40 m and a brief duration of a few minutes [18]. This is because, given the sharpness of the fingerprints that are observed on the innermost side, the Sudarium had to be opened very briefly after the inner stains were produced.

In the second Congress of 2007, Heras[19] claimed to have found the print of the right ear on the right half of the Sudarium. It implies the deployment of the Sudarium around the whole head.

4.4.2 Accordion and Diffuse Stains

We would like to analyse here the bellows or accordion-shaped stain located to the right of the observer. Its probable genesis is a stain similar in composition to the main stains. Their pointed edges correspond to the cloth folded over itself like a tube and then crushed before the fabric had dried. If we look comparatively at the upper arched edges, we can verify the agreement of multiple points that are repeated by impregnation. The limiting edge is well fixed at the top but is more diffuse at the bottom. A new experimental series allowed us to verify that this occurs when heterogeneous pressures are applied to the folded fabric, limiting it more precisely when higher pressures are applied and diffusing slightly in the areas where this pressure is lower. Consequently, this stain originated in a secondary way with

[18]Heras, G., et al. (1993). *El Santo Sudario de Oviedo.* Proceeding of the 2nd National Congress of Paleopathology (Valencia, October 1993). p. 354.

[19]Heras, G., and Ordeig, M. (2007). *Consideraciones geométricas sobre la formación central de manchas del Sudario de Oviedo.* Proceeding of the 2nd National Congress on the Sudarium of Oviedo. pp. 237–266.

respect to the main stains and from these, due to a folding of the outer fabric, being subjected to a greater pressure at the top and almost null at the bottom.

The reconstruction is therefore that, when the cloth was still folded with a double layer in front of the face, the excess part was folded in three like a tube on the left part of the beard and fastened by stitches.

For the diffuse stain that is between the pointed stains and the main stain of the left, the best simulation was obtained when a stained cotton swab was placed on the linen cloth. The similarity was complete when stained hair was placed on the cloth and a diluted blood solution was added. This long and complex experimental series specified that this intermediate diffuse stain was caused by hair, stained, and conglutinated with blood, on which a diluted blood solution slowly flowed. This part of the Sudarium was in fact over the lock of hair of the left side of the head of the deceased.

4.4.3   Pointed Stains

The pointed form or dot-shaped stains near the left edge, corresponding to the occipital and nape area could be the wounds from the crown of thorns. They could not be reproduced in the test, using blood from a corpse nor from the blood bank of the Hospital which included an anticoagulant. In those cases, the resultant stains lacked the concentric halos around a denser nucleus that are visible in the pointed stains of the Sudarium.

The Villalaín team then proceeded to experiment with whole blood, coming directly from the experimenter himself, without additives. Thus, they were able to verify how the coagulation process formed the aforementioned concentric halos and that these varied as a function of the time elapsed between the blood flow and the moment of impregnation on the linen. In view of the experimental series, it can be stated that the pointed stains are caused by vital blood, dripping from numerous dot-shaped bleeding wounds and that the cloth was applied on the wounds, approximately 60 minutes after they bled (Fig. 4.19).

This means that the wounds from the crown of thorns were bleeding for about an hour before the use of the Sudarium on the head.

Figure 4.19 Simulation of the pointed stains with vital blood (© CES).

The different nature of these pointed stains when compared to the main stains is also confirmed by the infrared and ultraviolet photographs.

4.4.4 Butterfly Stain

The butterfly stain is just below the spots of vital blood from the nape, but it consists of post-mortem blood. The different nature of both fluids is manifested by infrared illumination. The butterfly stain disappears and the stains on the nape do not disappear under infrared illumination (see Fig. 4.20). The lack of infrared emissivity is a characteristic that the butterfly stain shares with the main stains, which suggests a common origin. The lighter tonality also connects this stain with the main stains.

The symmetry of the double wing shape of the stain suggests that it originates from when this part of the Sudarium was applied around a stained structure. From the experiments carried out, this structure should be hair and thus it was possible to reproduce a similar stain, applying linen cloth around a cotton bundle (Fig. 4.21). The stain itself and the surrounding wrinkles allow us to suspect the existence of hair gathered in the form of a "ponytail", stained with diluted blood.

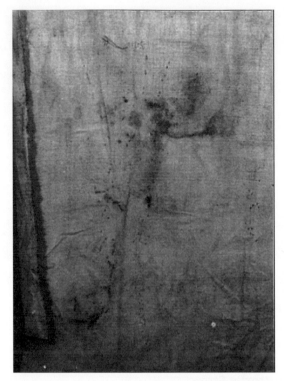

Figure 4.20 Infrared image of the butterfly stain (© CES).

Figure 4.21 Simulation of the butterfly stain (© CES).

4.4.5 Corner Stain

The corner stain is close to the butterfly stain and has the similar characteristic and behaviour under infrared light as it has (Fig. 4.20). Dr Sánchez Hermosilla and other members of the EDICES claim that this stain could have its origin in a back wound where the spear exited the crucified.

Although they search for confirmation from the image of the Shroud of Turin, there is no evidence of such an exit wound. Moreover, when the pointed stains are superimposed over the similar group in the nape of the Shroud, the corner of the Sudarium falls to the back of the neck near the cervical vertebrae. This position is too high to correspond with a wound from the spear coming out of the back of the Man of the Shroud (Fig. 4.22).

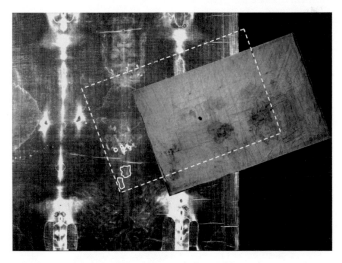

Figure 4.22 Superimposed contour of the Sudarium over the wounds of the nape of the Shroud of Turin and place of the corner stain (red).

A possible alternative was proposed in the 2nd Congress of the Sudarium in 2007[20]. If the nature of the corner and butterfly stains is the same as of the main stains, the simplest hypothesis is that both have the same origin, that is; the flow from the nose

[20]Barta, C. (2007). *Aproximación del Equipo de Investigación del Centro Español de Sindonología al estudio comparativo Sudario de Oviedo-Síndone de Turín.* Proceeding of the 2nd National Congress on the Sudarium of Oviedo. pp. 393–424.

and mouth. In fact, a small stream of blood comes out of the right corner of the mouth, leaving its mark on both symmetric main stains (Fig. 4.23). This stream reaches the fold of symmetry and goes away beyond the part covered by the cloth. Then, it could run along the uncovered part of the neck (Fig. 4.24).

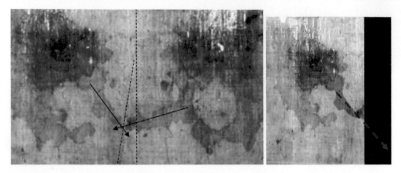

Figure 4.23 Blood stream from the mouth in both symmetrical stains (left). The stream reaches the fold of symmetry and could flow beyond (right).

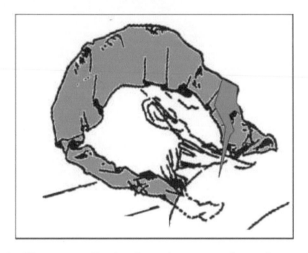

Figure 4.24 The stream flowing from the nose and mouth runs beyond the Sudarium reaching the neck.

The flow from the nose and mouth can reach the back of the neck as was verified through a simple simulation with a sculpture of the Man of the Shroud at half size (Fig. 4.25). A small quantity of gel was dropped on the base of the nose and left to run purely by the action of gravity. We observe that the stream reaches the back of the neck at just the same level where the butterfly and corner stains appear.

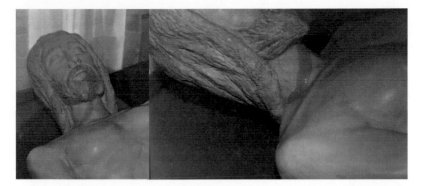

Figure 4.25 The stream flowing from the nose and mouth runs beyond the Sudarium reaching the back of the neck.

The only condition was that the corpse was face up for some time when the cloth was still folded with double layer. This position is included in the hypothesis formulated by Heras and Villalaín as we saw above. The corpse must have been translated face up for a short path and for a brief duration. During this manipulation, the flow tends to increase due to movement and, following the gravity, reaches the back of the neck in sufficient quantity to form major stains that soak the hair that was rolled through stitches inside the Sudarium.

This hypothesis for the formation of the butterfly and corner stains is coherent with all the previous findings on the stain characteristics.

4.5 Summary

It was proved by three independent specialists that the stains of the Sudarium are human blood and most probably of the group AB. The human nature of the blood was confirmed by the identification of mitochondrial DNA. The reliability of these statements is limited by the doubts of possible contamination always present.

The presence of dried blobs of fibrin supports the assumption that the Man of the Sudarium had undergone a severe trauma such as scourging some hours before the use of the Sudarium on him.

The main stains observed in the Sudarium of Oviedo were produced by a mixture of blood and pulmonary edema in the rate of 1/6 that flowed slowly from the nose and mouth while the corpse was in vertical position with the head inclined 70 degrees forward in relation to the vertical axis. This blood has been determined to be post-mortem blood. The Sudarium was folded in order to have two layers in front of the face. The hypothesis of a crucifixion in a Jewish context justifies the goal of saving the blood of the corpse in order to bury it with the deceased.

The Sudarium, therefore, would have to have been placed on the head of a crucified man who had died about an hour before. With the onset of rigor mortis, the pulmonary edema would begin to flow, soaking the beard and moustache. In this position, the lower parts of the main stains were formed.

After approximately one hour, the body would then be taken down with the arms still fixed to the horizontal beam and placed face down, with pulmonary liquid continuing to flow from the nose, thus forming the stain that appears in the area of the forehead. In this position it remained for about one more hour.

At some point the corpse was transported face up and the flow from the nose was able to reach the nape.

During a final phase, the Sudarium had to be opened up to surround the whole head.

In the nape area, the pointed stains indicate that the wounds were inflicted in life and they were bleeding until about an hour before the placing of the Sudarium on the head.

The correspondence between the stains and the anatomical features enables the measuring of the head and face of the man of the Sudarium. However, this subject will be analysed later.

Chapter 5

The Calvary Sequence Undergone by the Corpse

5.1   Sequency

With all the data analysed in the previous chapters we can now reconstruct the final hours of the Man of the Sudarium. We have determined how the Sudarium was fixed to the hair, how it was folded, what was the tilt of the head at different times, the origin of the different blood stains, how it was sprinkled with aromas, etc. With all these data, we can propose a sequence that matches all of them. We realize it may have some incorrect assumptions, but it allows us to construct the basic sequence of the last moments in the life of the Man of the Sudarium.

First of all, the Man of the Sudarium underwent a traumatic experience before the use of the Sudarium on him. The presence of fibrin was the consequence of a severe trauma suffered some hours prior that could have been the scourging. This punishment also explains the pulmonary oedema present on the Man of the Sudarium. The effort to carry a weight, such as the cross or a part of it, could also contribute or be the traumatic experience. This oedema is proven by the nature of the stains in the Sudarium.

The Sudarium of Oviedo: Signs of Jesus Christ's Death
César Barta
Copyright © 2022 Jenny Stanford Publishing Pte. Ltd.
ISBN 978-981-4968-13-3 (Hardcover), 978-1-003-27752-1 (eBook)
www.jennystanford.com

The Man of the Sudarium was face down to earth for some time before the use of the Sudarium as is indicated by the limestone dust close to the tip of the nose. As a suggestive hypothesis, if He had his hands attached to the beam during the ascent to the Calvary and He fell, He necessarily hit the ground with his face.

The head of the Man was injured in life by dot shaped objects which fits perfectly with a crown of thorns. This crown was on the head until the placing of the Sudarium. An hour before, the wounds of the nape were bleeding as it can be concluded because of the nature of the associated clots as we saw in the previous chapter. Then, the Sudarium was placed on the head about an hour after the death of the Man.

Arbitrarily, we assume that the death happened at 3:00 p.m. The deceased Man had undergone a traumatic punishment and had a crown of pointed objects in the head (Fig. 5.1). He was in a vertical position compatible with the cross.

Figure 5.1 3:00 p.m. Death of the Man of the Sudarium. Sculpture created by' by Dr Juan Manuel Miñarro.

Almost an hour after the death, at about 4:00 p.m., the nose and the mouth of the Man started to slowly leak a reddish fluid like blood soaking the beard and moustache (Fig. 5.2). This proves that the Man did not breathe and was already a corpse. If the friends of the Man were Jewish, they would have feared that the blood was going to be lost to the earth. They had to prevent this loss because the blood had to be buried with the corpse according to their beliefs.

Figure 5.2 4:00 p.m. Bleeding by the nose and mouth.

They decide to remove the crown (Fig. 5.3) and to place the Sudarium around the head and in front of the face to soak the cloth and restrain the blood flow.

Figure 5.3 4:00 p.m. Crown of thorns removed (© CES).

The head of the corpse was inclined 70 degrees forward, allowing access to the nape. Someone placed the edge of the Sudarium along the nape and wrapped a lock of hair with it and he (probably a man if it was a Jewish context) sewed the cloth to the hair to fix the Sudarium tightly.

Figure 5.4 4:00 p.m. An edge of the Sudarium is sewn to a lock of hair at the nape.

About half of the rest of the cloth was passed around the left ear and in front of the face, covering it and reaching the right cheek. There, the cloth was folded to turn back towards the face providing a second layer in front of the mouth and nose area (Fig. 5.5).

The fold over the right cheek was sewn to the beard and to the hair on this side of the face and the other free edge was wrinkled on the left side and fixed in some fashion. In this position, the Sudarium remained for about an hour soaking the bloody fluid and producing the lower part of the symmetric main stains in front of the nose and mouth (Fig. 5.6).

After approximately one hour, at about 5:00 p.m., once the Roman centurion pierced the corpse's thorax, the body was taken down probably with the arms still fixed to the horizontal beam[1]. For unknown reasons, the caretakers who moved the

---

[1]Hypothesis of how the Man of the Shroud was manipulated can be found in Bevilacqua, M. et al. (2014). How was the Turin Shroud Man crucified? Injury. Barbet P. (1963). *A doctor at Calvary: the passion of our Lord Jesus Christ as described by a surgeon.* In: Earl of Wicklow (trans). Garden City, NY. Image Books Ed.

body decided to turn the corpse face down (Fig. 5.7). If the blood was dripping by the nape, they could think that the two layers covering the face would retain the blood more effectively. Or they chose it because of modesty. We do not have a proven hypothesis.

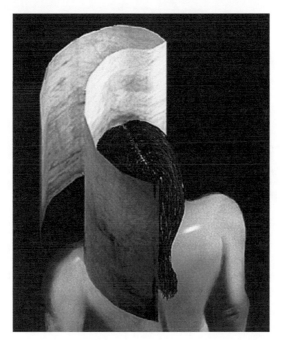

Figure 5.5 4:00 p.m. The Sudarium is passed over the left ear and covers the face (© CES).

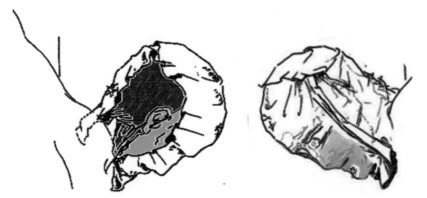

Figure 5.6 4:00 p.m. Position for the first stage.

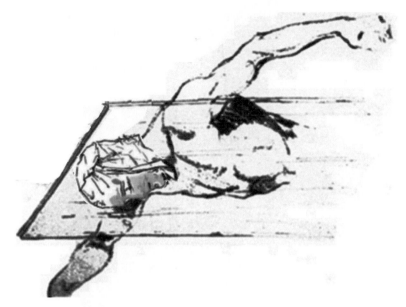

Figure 5.7 5:00 p.m. Position face down for the second stage (© CES).

In this posture, the nasal flow ran into the back of the nose and soaked the upper part of the main stains that were covering the bridge of the nose and the forehead. The corpse remained like this for almost another hour.

Afterwards, the corpse was placed face up, still remaining with two layers in front of the face and his arms were placed along the sides of his body. This movement resulted in an increase of blood leakage from the nose and mouth. Probably at this moment, this flow reached the back of the neck and soaked the corner of the Sudarium fixed over the nape producing the "butterfly" stain and the corner stain (Fig. 5.8).

At about 6:00 p.m., for a few minutes the body was waggled, probably to transfer it some distance because, the stains that soaked the cloth at that time correspond to a brief but intense flow. Probably a hand on the nose tried to block the exit of more blood.

The position of the Sudarium was altered leaving only a layer covering the face, wrapped around the whole head of the Man. A knot was tied at the top of the head (Fig. 5.9).

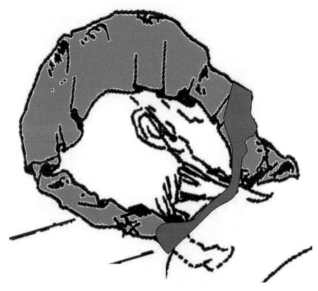

Figure 5.8 6:00 p.m. To transport the corpse, He was placed face up for a short period.

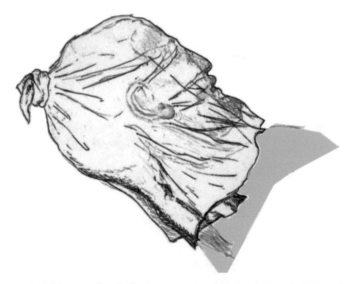

Figure 5.9 6:00 p.m. The Sudarium wrapped around the whole head with a knot in the top.

Finally, the Sudarium had to be removed from the corpse because it was sprinkled with aloes which are still imbibed in the blood and remain on the surface that had been in contact with the face. It implies that the Sudarium was no longer wrapping the head.

We can assume that the Sudarium with the blood of the deceased Man had to be buried near Him in the sepulchre if the burial was a Jewish entombment.

In this case, the corpse had to be covered with a Shroud and the Sudarium that had covered his head, remained not with the burial cloths but rolled up in a separate place (Fig. 5.10 and Fig. 5.11).

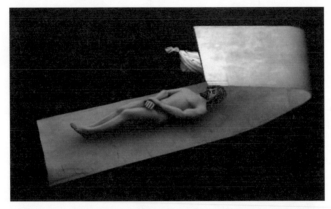

Figure 5.10 6:10 p.m. The Sudarium was removed, and the deceased was covered with a shroud.

Figure 5.11 6:10 PM. The Sudarium remained near the deceased until somebody recovered it.

5.2 Summary

The Man of the Sudarium underwent traumatic experiences before the use of the Sudarium on him: a punishment that involves pulmonary oedema and wounds around the head produced by dot shaped objects (crown of thorns?).

We can reconstruct the final hours of the Man of the Sudarium. We arbitrarily assume that his death happened at 3:00 p.m. The Sudarium was placed around the Man's head almost an hour after his death, around 4:00 p.m., when He was still on the cross. He remained in a vertical position with the Sudarium for about another hour. Around 5:00 p.m., the body was taken down and the corpse was left face down. The corpse remained like this for almost another hour.

Afterwards, around 6:00 p.m., the corpse was placed face up, and moved to the tomb. Just before the corpse was covered with the shroud, the Sudarium was removed and put aside close to the corpse in the sepulcher.

Chapter 6

Dating of the Sudarium

In this chapter several dating tests by the carbon-14 (C14) method of samples of the Sudarium will be presented. We will analyse the reliability of the method and possible contradictions and errors which could invalidate the result of the C14 dating. Finally, some alternative methods for dating will be proposed.

The Sudarium of Oviedo was dated with the C14 method. Isotope 14 of carbon is continuously produced in the upper layers of the atmosphere when nitrogen is bombarded by neutrons (n) produced by cosmic rays.

$$_{0}^{1}n + {}_{7}^{14}N \rightarrow {}_{6}^{14}C + {}_{1}^{1}p \tag{6.1}$$

In this reaction, a proton (p) is emitted. As this isotope of carbon, C14, is radioactive with an established half-life of 5730 years, it disappears from the atmosphere when it decays according to the following reaction:

$$_{6}^{14}C \rightarrow {}_{7}^{14}N + \beta^{-} + \overline{\nu}_{e} \tag{6.2}$$

Where the beta particle β^{-} is an electron emitted by the nucleus and ν_{e} is a neutrino associated to the disintegration of

The Sudarium of Oviedo: Signs of Jesus Christ's Death
César Barta
Copyright © 2022 Jenny Stanford Publishing Pte. Ltd.
ISBN 978-981-4968-13-3 (Hardcover), 978-1-003-27752-1 (eBook)
www.jennystanford.com

neutrons. A balance is reached between what is created and what disappears, resulting in a fairly constant amount of C14 in the atmosphere. Every plant such as flax (linen), breathes CO_2 and incorporates C14 into its tissues with a ratio of the radioactive isotope close to that found in the atmosphere. Other living beings in the animal kingdom incorporate this proportion through the food chain. But when they die or, in the case of crops, are harvested, the amount of C14 is no longer renewed and begins to decline. The death of the living being can be considered as zero time. At that moment, the ratio of C14 roughly corresponds to that in the atmosphere. As time passes since death, the ratio of C14 decreases. By measuring the content of C14, one can therefore determine the time when the animal died, or the plant was collected. This is the principle of C14 dating that is currently applied to small samples by measuring their content of that isotope with a mass accelerator spectrometer (AMS). The results are given by convention as years Before Present (BP) and fixing the "Present" as 1950 AD (*Annus Domini* = Year of the Lord).

The raw C14 age is converted to calendar date through the calibration curve (see Section 6.2) built with the dendrochronology or the tree rings dating. Before processing the sample, the laboratory applies cleaning treatment to eliminate possible contamination.

6.1 First Dating Attempts

Among all the dating techniques, the described one that uses the radioactive decay of the C14 isotope is the most popular. As a technique based on physical-chemical principles, it is applicable to fabrics such as the Sudarium of Oviedo.

Msgr Ricci had shown interest in applying this dating technique shortly after starting his investigation and requested samples and permission to the owner in Oviedo to proceed. But in a first attempt there were no results. The following is a summary of the intricate story about the first dating attempt that led to some confusion[1], which was finally clarified with the following explanations[2]:

- Around 1977, Msgr Ricci cut out and sent to McCrone a sample of the Sudarium.
- McCrone sent it to the Lawrence Berkeley Laboratory
- Dr Richard Muller transformed it into CO_2 and sent it to the AMS Laboratory in Tucson (Arizona, USA).
- Paul Damon, director of the laboratory, received it in gaseous state in an unsealed vial that rendered it useless.
- No dating is done.
- Damon reports this to Mark Guscin (letters of 1995).

There is nothing more to be said about this unsuccessful dating attempt and we can forget it.

Msgr Ricci continued to bear in mind the intention of dating in his investigation. He requested new samples from the Council of canons of Oviedo to date them using by carbon-14 technique.

In 1990, the Sudarium of Oviedo was finally dated by the carbon-14 method at the request of the Italian researcher Mario Moroni, a member of the International Center for Sindonology in Turin (CIS) when Dr Baima Bollone was the president. At the Oviedo Congress in 1994, the result was communicated and recorded in the proceedings[3]. Further details were published at the Orvieto Congress in 2000[4] and in the *Collegamento Pro Sindone*[5] Magazine.

Two samples of the Sudarium were dated which, according to the above references, were cut out by Frei in 1979. Macrophotographs of these samples sent to Arizona and Toronto are shown in Fig. 6.1.

[1]Guscin, M. (1999). *El Sudario de Oviedo y el Carbono 14*. Linteum 24–25 December 1998 Mars 1999. pp. 15–18.

[2]Moroni, M., et al. (2000). *A Suggestive Hypothesis covering the radiodating results of the Sudarium of Oviedo and the Shroud of Turin*. The Orvieto Worldwide Conference "Sindone 2000". August 27–28–29, 2000.

[3]Bollone, B., et al. (1994). *Resultado de la valoración de las observaciones y de los exámenes de algunas pruebas efectuadas sobre el Sudario de Oviedo*. Proceeding of the 1st International Congress of the Sudarium of Oviedo. pp. 428–431.

[4]Moroni, M., et al. (2000). *A Suggestive Hypothesis covering the radiodating results of the Sudarium of Oviedo and the Shroud of Turin*. Congress de Orvieto 2000.

[5]Moroni, M., (1999). *La Radiodatazione del Sudario di Oviedo*. in Collegamento Pro Sindone n.3. Mars-April 1999. pp. 29–36.

Figure 6.1 Samples dated by the Moroni team.

The results are indicated in Table 6.1[6]. The content of C14 obtained by the two laboratories corresponds to a date later than the second half of the 7th century.

Table 6.1 Previous dating of two samples of the Sudarium

Sample	Sudarium 1	Sudarium 2
Material	Linen	Linen
Weight	20.79 mg	14.00 mg
Origin[a]	Sample taken by Max Frei between 15–17 May 1979	Sample taken by Max Frei between 15–17 May 1979
Expected date	First century AD	First century AD
Cleaning Treatment	Acid-Alkaline-Acid	Acid-Alkaline-Acid
Laboratory	Arizona	Toronto
Analysis date	31 October 1990	11 September 1991
Laboratory identification	V6009 (AA6049)	MM60 (TO2442)
Method	AMS	AMS
Raw result	1292 ± 53 years BP	1300 ± 40 years BP
Interval of dates (%confidence.)	642 – 869 A.D. (95%)	653 – 786 A.D. (95.5%)
δ13C	–25%	–25%

[a]Although the report of Moroni indicates the date of May 1979, the minutes of the Council of Canons of the Cathedral of Oviedo recorded the date of sample taking the 3rd August 1977. López, E. (2004). El Santo Sudario de Oviedo. Ediciones MADU. Asturias 2004. p. 89. nota 99.

To allow anonymity respect to the Sudarium in the process, the laboratory certificate was managed by Giuseppina Valsecchi (Mario Moroni's wife).

---

[6]Moroni, M., et al. (2000). *A Suggestive Hypothesis covering the radiodating results of the Sudarium of Oviedo and the Shroud of Turin.* Congress de Orvieto 2000.

Another sample, which had previously been scorched at 190°C for 30 minutes, was also dated[7] and the results were presented at the Scientific Symposium in Rome in June 1993. A sample of the reinforcing fabric was also dated (Fig. 6.2). It is the cloth that, was most probably ordered by Diego Aponte de Quiñones (1585–1598) and given to Msgr Ricci by the Council of canons in 1978 (or 1977). These two results are shown in Table 6.2.

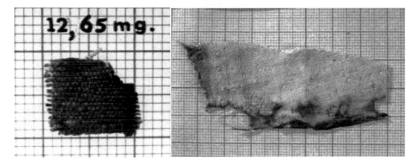

Figure 6.2 Scorched sample and support cloth dated by the Moroni team.

Table 6.2 Dating of scorched sample and support cloth

Sample	Interval of dates (% confidence)
Sudarium 3 (Scorched)	540 – 754 A.D. (95%)
Support Fabric	1471 – 1679 A.D. (90.1%)
	1765 – 1803 A.D. (6.7 %)

The purpose of this experiment was to show that a fire, like the one suffered by the Shroud of Turin in 1532, makes the fabric look younger. The fire explanation for the radiocarbon dating of the Shroud of Turin had been supported by Dr Kouznetsov[8]. However, the heat treatment to a sample of the Sudarium gave a result that does not support this explanation. As we can see, the heat modified the date only slightly with respect to the unheated sample, but in reverse: the heat aged the sample by about 100 calendar years.

---

[7]Moroni, M., et al. (2000). *A Suggestive Hypothesis covering the radiodating results of the Sudarium of Oviedo and the Shroud of Turin.* Congress de Orvieto 2000.
[8]Kouznetsov, D., et al. (1996). *Effects of fires and biofractionation of carbon isotopes on results of radiocarbon dating of old textiles: The Shroud of Turin.* Journal of Archaeoloqical Science. 23, pp. 109–121.

The other sample, the support cloth, is from the late 16th century according to the Diocesan Center of Rome. The dating of the support cloth gives a date that is perfectly compatible with that expected. This would imply conditions that during the time in which both fabrics were united the environmental have not modified the content of C14. This test also allows assessing the possibility of isotopic exchange when two cloths are in close contact for centuries. That is: the Sudarium could get younger because new C14 passed from the support cloth to the Sudarium and vice versa. In a first approach, the number of centuries in which the Sudarium would reduce its age would be the number of centuries in which the support cloth would increase its age. The results of the C14 dating of the support cloth invalidated this hypothesis because it would have to be modern and would only increase its age by four centuries, while the Sudarium would reduce its age by six/seven centuries.

Table 6.3 indicates additional details.

Table 6.3 Data of scorched sample and support cloth

Sample	Sudarium 3	Support cloth
Material	Linen	Linen
Weight	13.00 mg	22.00 mg
Origin	Tissue removed by Max Frei between 15–17 May 1979[a]	Delivered to M. Moroni by Msgr Ricci on 5 February 1994
Expected date	First century AD	Diego Aponte de Quiñones 1585–1598
Cleaning Treatment	Acid-Alkaline-Acid	Acid-Alkaline-Acid
Laboratory	Arizona	Zurich
Analysis date	7 September 1992	19 May 1994
Laboratory identification	(AA8432)	ETH 12008
method	AMS	AMS
Raw result	1450 ± ?[b] years BP	285 ± 50 years BP
$\delta13C$	−25.6%	

[a]About the sample taking see above note 7.

[b]The uncertainty has not been reported by Moroni in the cited reference but should not differ significantly from that assigned to the other samples (±.60 years).

Regarding the dating carried out by Mario Moroni, there are certain reservations. In the communication of the 1994 Oviedo congress[9], the authors already indicated the uncertainty about the conditions of conservation of the samples since they were taken (1979) until they were in the hands of the International Center of Sindonology of Turin (CIST), years after the death of Max Frei which occurred in 1983.

The information on the collection of these samples from the Sudarium has not been published in full detail and is not sufficiently clear. Where were the dated samples taken from and by whom? Baima Bollone in the 1994 Congress[10] says that it was Frei in 1979 who obtained the samples. However, Enrique López[11] assures that the samples were taken on August 3, 1977, by Ricci and that they were 3 *"from the upper right. Previously, it had been tried from the front, lifting the silver gallon that delimited the Sudarium in contact with the frame, but it was given up because the silver gallon was sewn to the Sudarium itself from behind. The measurements of the cut pieces were: one of 9 cm long, another of 5 cm and the third of 4 cm. The width was always less than the gallon"*. A sample was also burned by McCrone in the first failed dating attempt. The lengths of the samples dated by Moroni's team are about 1.7, 1.2 and 0.9 cm, giving a total of less than 4 cm. E. López also tells us that Frei did not take any cuts. He received the samples from Ricci. The samples found in the Frei archive were sent to the International Center for Sindonology. Finally, in 1990 Baima Bollone gave to Mario Moroni the three samples of the Sudarium from the Frei archive. It has been confirmed that there is no trace of any remaining sample either at the *Centro Diocesano di Sindonologia* "Giulio Ricci" or at the *Centro International de la Sindone* in Turin.

It seems that the dated part was in intimate contact with the silver gallon. When in Oviedo they said *"upper right"* it is most likely that they were referring to the way in which the bishops

[9]Bollone, B., et al. (1994). *Resultado de la valoración de las observaciones y de los exámenes de algunas pruebas efectuadas sobre el Sudario de Oviedo*. Proceeding of the 1st International Congress of the Sudarium of Oviedo. pp. 428–431.

[10]Bollone, B., et al. (1994). *Resultado de la valoración de las observaciones y de los exámenes de algunas pruebas efectuadas sobre el Sudario de Oviedo*. Proceeding of the 1st International Congress of the Sudarium of Oviedo. pp. 428–431.

[11]López, E. (2004). *El Santo Sudario de Oviedo*. Ediciones MADU. Asturias. p. 89, nota 99.

traditionally exposed the Sudarium. However, this does not define what the *upper right* is because it depends on the date when the Sudarium was exposed. As we said in a previous chapter, the Sudarium was shown on one side and the opposite after some intervention and the way it was fixed to the frame also changed. Figure 6.3 shows two different ways the Sudarium was displayed and in both there is a cut in the upper right corner, although it is hardly discernible in exposition on the left.

Figure 6.3 Orientation of the Sudarium during old expositions (© CES).

Thanks to subsequent analysis[12], it was concluded that the samples for the C14 dating should have been cut from the corner adjacent to the tear (upper side in Fig. 6.3 left).

Therefore, they would be referring to the area adjacent to the tear. Furthermore, in 1986 Dr Saavedra also cut 4 cm divided into two fragments for Ricci. D. Enrique consulted the Benedictine MMs of the San Pelayo Monastery for this purpose, who had modified it in 1987 by removing the silver gallon. They confirm that the top strip near the tear was already missing. The three samples mentioned in the 1977 minutes must have been taken from this area (Fig. 6.4). The additional sample to be dated under the management of the EDICES was also cut from the same place and it is also shown in Fig. 6.4. The 1986 samples were probably removed from the Ricci corner.

---

[12]Barta, C. (2007). *Datación radiocarbónica del Sudario de Oviedo*. Proceedings of the 2nd International Congress for the Sudarium of Oviedo. pp. 137–158.

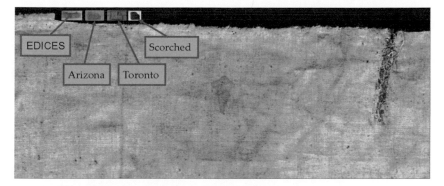

Figure 6.4 Location of the four samples used for C14 dating.

The weights per unit area that are deduced from the photographs and the tables of data are slightly lower than the weights per unit area measured by the EDICES. The discrepancy may be related to the ability of the linen to absorb moisture when the measurement of the whole Sudarium was made in the cathedral of Oviedo.

Even if the traceability and conservation conditions are not those required, we can give some credibility to the results of the dating carried out by the Italian group.

6.2 Dating Managed by the EDICES

To clarify any doubts about the traceability of the dated samples, the EDICES repeated the dating by cutting out a scrap adjacent to the area from which the previous samples were cut for dating. The sample was apparently clean according to the microscope observation by Felipe Montero (see Fig. 6.5). The tissue was sent to the Beta Analytic Laboratory and the result published on 13 February 2007 confirms the previous results. The details of the result are shown in Table 6.4.

As we said in the introduction to this chapter, the raw C14 age is converted to calendar date through the calibration curve (Fig. 6.6). The isotopic C14 direct measurement has a Gaussian distribution of probability (Y axis). When it is projected against the calibration curve, we obtain the probability for the date of the harvest of the linen (X axis). In the case of the Sudarium of Oviedo, the maximum probability of the raw data projects a maximum probability of calendar age between 748 AD and 800 AD. The central date of the 95% of confidence level for the calendar interval is also 750 AD. This could be the reference date for a coarse result of the C14 dating.

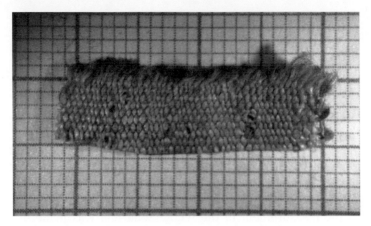

Figure 6.5 EDICES sample used for C14 dating (© Montero and CES).

Table 6.4 Dating of the EDICES sample of the Sudarium

Sample	Sudario 4 (SO-1711)
Material	Linen
Weight	22.8 mg
Origin	Sample taken by Felipe Montero on 17 November 2006
Expected date	First century AD
Cleaning Treatment	Acid-Alkaline-Acid-Cellulose extraction
Laboratory	Beta Analytic
Analysis date	13 February 2007
Laboratory identification	Beta – 225639
method	AMS
Raw result	1240 ± 50 years BP
Interval of dates (% confidence.)	660 – 890 A.D. (95%)
$\delta 13C$	–25.2%

If the result is accepted as valid, then the authenticity of the relic would have to be rejected. However, this is not consistent with the rest of the conclusions from the other studies conducted on the Sudarium.

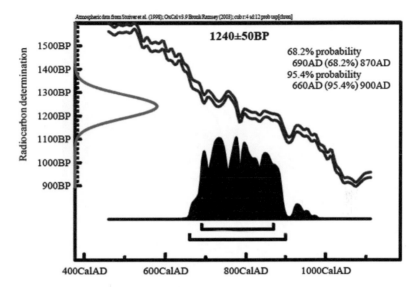

Figure 6.6 Calibration curve to correlate raw data to calendar date.

The forgery seems quite unlikely because the absence of details in the Gospel of John prevents the hypothetical forger from imagining the sequence of use deduced from the forensic analysis. In fact, some scripture specialist interpreted that the Sudarium of the Gospel to be a chin band or a headdress. In the artistic representation of tomb after the Resurrection, the type of sudarium that the artist imagined was also something as a headdress (Fig. 6.7). A forger would have made this other type of clothing. However, there is no chin band that is claimed to be the sudarium of the Gospel.

Other reason to rule out forgery is the difficulty to understand what the Sudarium of Oviedo exhibits. A forger had to create something recognizable to his contemporaries; otherwise, the forgery would not be successful. However, for centuries, ecclesiastics who displayed the Sudarium to the faithful have presented it in an upright position (see Fig. 6.3). It means that they did not realize that the Sudarium had the outlines of a face that they rotated 90° in a strange way when they displayed the cloth (Fig. 6.8).

Figure 6.7 Resurrection. Constantinople 12th century (© Louvre).

Therefore, we do not think that the results of the C14 dating are sufficient reason to refute the tradition without considering the reasons that support its authenticity. In addition, the C14 method has its limitations as we will see in the following section and there are other indications and documents that place the Sudarium of Oviedo more likely at a date prior to that provided by the radiocarbon test.

Figure 6.8 The way the bishops displayed the Sudarium is evidence that they did not understand that the relic enclosed the outlines of a face.

6.3 Uncertainties of C14 Method

The C14 dating method often gives brilliant results but has limited reliability. Although the theoretical principles are well founded, when applied in real conditions, a non-negligible percentage of measurements provide aberrant results that archaeologists consider invalid. They compare the results of C14 with other criteria such as stratigraphic sequence of the soil, the cultural context or additional archaeological material.

We have carried out a study on 113 dating of samples of a varied nature (wood, bone, etc.) published in the specialized journal Radiocarbon[13] with the purpose of estimating the success rate of C14 dating. The archaeologists, when filling in the dossier to be sent to the laboratory with the information on the sample to be dated, usually give very generic indications on the chronology of the deposit so as not to condition the result. That is to say, for example, they do not state that they expect approximately

---

[13]Boskovic, R. (2002). Institute Radiocarbon Measurements XV. Radiocarbon Volume 44. Number 2.

16,500 BP, but rather they indicate Upper Paleolithic (between 35,000 and 10,000 BP). A vast majority of the measurements give results that confirm what was expected but, despite the fact that the archaeologists' way of proceeding avoids discrepancies between the laboratory dating and the proposed date, more than 20% are more than 2σ (sigma or standard deviation) of the expected date (See Table 6.5)[14].

Table 6.5 Reliability of the C14 dating method against the expected

Deviation from the expected	Reliable	Doubtful	Unacceptable
% of samples	76%	7%	17%
Number of samples	86	8	19

Reliable = deviation within the interval $[-2\sigma, +2\sigma]$
Doubtful = deviation within the intervals $[-3\sigma, -2\sigma]$ and $[+2\sigma, +3\sigma]$
Unacceptable = deviation less than -3σ or greater $+ 3\sigma$

A normal distribution of the results would have contained 95.4% of the cases in the interval $[-2\sigma, +2\sigma]$ and 99.7% of them in the interval $[-3\sigma, +3\sigma]$. Outside this last interval, a percentage of around 0.3% would remain. However, the distribution of the results found for the study of the 113 datings shows the particularity of containing more cases for deviations greater than $\pm3\sigma$ than those that occur between $\pm2\sigma$ and $\pm3\sigma$, which indicates that when there is a discrepancy, this is very significant. Among the most important deviations there are 20, 30 or 96σ. It is also observed that within the acceptable results there is a slight bias towards younger dates than expected (Fig. 6.9).

For sure, the reliability percentages obtained in this study cannot be considered representative of the method in general, since they have been carried out with datings in a single laboratory that receives samples from a very localized area in Central Europe. No information is provided on the cause of the discrepancies between what was expected and what was obtained by C14 dating. It is even possible that the archaeologist assumed the wrong dating and that he recognized his error in

---

[14]For some additional data, see Barta, C. (2007). Datación radiocarbónica del Sudario de Oviedo. Proceedings of the 2nd International Congress for the Sudarium of Oviedo. pp. 137–158.

cases of unexpected results. However, other studies show that this does not seem to be the rule.

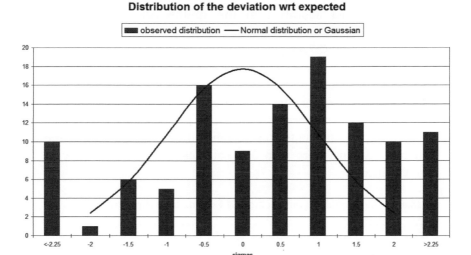

Figure 6.9 Deviation of dating wrt expected.

The percentages obtained by our particular study are close to those of rejection indicated by other archaeologists who show their value given to the C14 dating. This is, for example, the case of the archaeologist William Meacham[15] from Hong Kong, who provides his own percentages. Among the 115 excavations referred to by Dr Meacham, 78 (67.8%) were considered reliable, 11 (9.6%) dubious and 26 (22.6%) were rejected as unacceptable.

Another case that allows for the calculation of reliability percentages is that presented by Voruz and Manen[16], who classified 76 samples in the following categories and quantities:

<u>Reliable</u> (*retenues*) 44 (57.9%) are those that they consider as good or excellent and include them in the data base of reference dates.

---

[15]Meacham, W. (2000). *Thoughts on the Shroud 14C debate.* Sindon quaderno n°13, June 2000. pp. 441–454. Update in http://www.hku.hk/hkprehis/archdate.htm (2006) with similar percentages: among 125 samples, 88 (70.4%) were considered reliable, 10 (8.0%) were doubtful and 27 (21.6%) were rejected as unacceptable.

[16]Voruz, J. L., and Manen C. (2005). *Le Cas de la Grotte du Gardon (Ain)* Dossiers d'Archeologie n° 306, Septiembre 2005. pp. 38–43.

Acceptable 15 (19.7%) are those that are consistent with the expected time but that are accepted with some reserve. They are of secondary interest because a too large standard deviation, an unwarranted allocation to the stratum, or a shortage of contiguous objects for correlation.

Rejected 17 (22.4%) due to a result in deep disagreement with the type of adjacent furniture or due to inconsistency with the other dates of the same stratum. The causes of these failures are quite diverse: insufficient amount of carbon, contamination by older or more modern materials, confusion of layers, etc.

Table 6.6 presents the combined results of the three studies covering more than 300 samples and a wide variety of provenances.

Table 6.6 Combined reliability of the C14 dating from three studies

Reference	Reliable	Doubtful	Rejected	No. of samples
Barta	76%	7%	17%	113
Meacham	67.8%	9.6%	22.6%	115
Voruz	57.9%	19.7%	22.4%	76
Total	**68.4%**	**11.2%**	**20.5%**	**304**

These statistics show that, although the method of C14 dating is a useful method in archeology, it is far from infallible and is not a reason to invalidate an archaeological investigation when its result is at odds with the rest of the evidence.

The method was tuned for wood using the tree-rings chronology. Textiles, unlike trees, collect the C14 in a single year and are exposed to greater irregularity. In addition, once harvested and processed they are more permeable to the environment. Experience in textiles is more limited.

Ideally, an object selected for dating should have remained in the same environment that surrounded it since it ceased to be used. In that case, any possible contamination or exchange with the environment will incorporate C14 that is "aging" at the same rate as the object itself and will not disturb the dating. This is the case of archaeological sites, but it is not the case of the Sudarium of Oviedo and most of the relics that are frequently venerated in environments loaded by crowds and candles and are preserved along with other objects from different periods. As Montero remarked in a conference in the meeting of the CES in 2015, for centuries for the illumination of the Holy Chamber where the Sudarium was, oil of olives was used that produced smoke with modern C14 content.

However, there is today no solid scientific evidence that could really challenge the validity and accuracy of the radiocarbon datings that have been done for the Sudarium. There is not an identified cause that shifted the C14 dating from the first century to the eight centuries for the Oviedo's cloth. Some possible causes include that it suffered the consequences of the blasting of the Holy Chamber in 1934. Physical contamination with carbonaceous compounds, chemical reactions with strange carbonaceous substances, such as those promoted by micro-organisms, or radioactive transmutation of isotopes that consume or generate C14 are other possible causes that can lead to errors in the dating.

The EDICES did not consider the result of the dating sufficient reason to rule out the possibility that the Sudarium of Oviedo corresponds to what it is considered by tradition.

6.4　Incoherence with Other Data

The forensic analysis of the Sudarium of Oviedo points to a type of death that has a historical context. Consider the long list of death penalties used in ancient times[17]:

- Drowning
- Suffocation
- Crushing or pressing
- Gallows
- Crucifixion
- Decapitation
- Breaking wheel
- Burning
- Burial alive

[17]Von Hentig, H. (1967). *La Pena. Formas Primitivas y Conexiones Histórico Culturales.* Volume I. Espasa Calpe S.A. Translation of the original German of 1954.

- Dismemberment
- Stoning
- Impalement
- And more…

As we saw in the previous chapter, the result of the forensic analysis indicates that the Sudarium of Oviedo was applied to a person executed who remained in an upright position for more than an hour after expiration. We note that a vertical corpse is only compatible with the gallows and crucifixion among those listed. However, the mild effluvium of pulmonary oedema through the anatomical airways precludes the penalty of hanging[18]. The only remaining possibility is crucifixion. In other words, in all probability the victim wrapped in the Sudarium of Oviedo has suffered death on the cross.

The penalty of crucifixion was used since ancient times and precedes that of hanging. It was used by Assyrians, Egyptians, Persians, Greeks, Carthaginians, and Romans. It was abolished from the public courts by Emperor Constantine in the 4th century throughout his empire, probably with the edict of 325 AD. The historian Sozomeno (who died around 450) records of this resolution: "Constantine passed a law abolishing the torture of the cross used by the Romans until then"[19]. And Theodosius I the Great also abolished this capital punishment of the private courts around 392. The latest news of a Roman crucifixion is from 314 AD[20]. Therefore, the date around 750 offered by the C14 method is incompatible with the indications of the crucifixion presented in the Sudarium of Oviedo.

But we have to point out that this incompatibility is with a Roman crucifixion. However, there were crucifixions later in the East. In 1597 the Japanese crucified numerous Christian missionaries in Nagasaki[21]. And at the beginning of the 19th century

---

[18]Impalement could also keep the victim vertical by inserting, for example, the stick through the rectum and through the body of the victim until it comes out through the shoulder. However, Impalement can also be excluded. Pulmonary edema is a consequence of heart failure that does not occur in impalement where internal bleeding predominates. (Personal communication from Dr Villalaín).

[19]Igartua, J. M. (1989). *La Sábana Santa es auténtica*. Mensajero. Bilbao. p. 156.

[20]Zaninotto, G. (2000). *The Shroud and the Roman Crucifixion: A historical Review*. Sindon n.13 (Turin 2000) p. 299.

[21]Wilson, I. (1999). *The Blood and the Shroud*. Orion London. p. 238.

this form of execution was still common in some Asian countries and especially in Japan[22]. Isolated cases still appear today in Africa and the Philippines. Nevertheless, all these cases do not correspond to the date predicted by the C14 dating analysis and therefore they can be excluded from the origin of the Sudarium. They are also out of the context with respect to the Sudarium of Oviedo, which appears in what we could call Western-Roman culture.

Finally, we also found a possibility of compatible crucifixion that cannot be ruled out. Islamic law apparently includes crucifixion and there was a resurgence of this type of execution of Christians by Moors in the 7th century during the Arab-Christian conflict in the Middle East[23]. And in Spain there were crucifixions in Al-Andalus at least in 749 and 813. Even, the Sufi al-Hallâj was condemned, imprisoned, flogged, had his hands and feet mutilated and then crucified and cremated in Baghdad on March 26, 922 AD[24]. Therefore, if we focus only on the type of execution, the possibility remains, that the Sudarium could belong to a crucifixion by Muslims who had already begun to practice this type of execution at the time of the year 750 in which the relic has been dated. Even though the references of these Muslins crucifixions are scarce, we can assume the crucifixion of a Christian by Muslims in the South of Spain or in Middle East in the second half of the 8th century. We can even assume that the mockery could include the crown of thorns traces of which are also present in the Sudarium. The rest of the features found in the Sudarium are more difficult to conciliate. First of all, it must be analysed if the use of a sudarium to cover the face of the executed may fit into the aforementioned hypothetical event. Maybe Christians did not need to bury the blood of the deceased and did not use aloe and storax to preserve the blood-stained cloths. On the other hand, we wonder if the Islamic executioner would allow access to the corpse. An

---

[22]Von Hentig, H. (1967). *La Pena. Formas Primitivas y Conexiones Histórico Culturales*. Volume I. Espasa Calpe S.A. Translation of the original German of 1954. pp. 285–286.

[23]Zugibe, F. (2000). *Forensic and Clinical Knowledge of the Practice of Crucifixion: A Forensic way of the Cross*. Sindon n.13 (Turin 2000) p. 235 and Gramaglia. Pier Angelo. (1978). *L'Uomo della Sindone non è Gesù Crist o*, Turin, 1978 Referred in Celier, O. (1992). *Le Signe du Linceul*. Cerf (Paris 1992). p. 158.

[24]Celier, O. (1992). *Le Signe du Linceul*. Cerf (París 1992) p.158.

investigation of the Islamic and Christian customs in the 8th century is required to be carried out to know for sure. The details of many of these crucifixions are unknown and when they reach us, they sometimes include notable irregularities such as that of a crucified face down in 850[25]. In any case, as we will see, there is in the Sudarium a possible trace of the Roman scourge tool, the *flagrum*, which rules out the Islamic crucifixion theory.

In short, the crucifixion is the execution that best fits the evidence on the Sudarium. This suggests a time before the 5th century incompatible with the C14 dating. However, only if we find some feature that relates the type of crucifixion to the Roman stile, we will be able to rule out a possible later crucifixion out of the Roman customs.

Regarding a possible incompatibility of the C14 dating with documented references of the Sudarium specifically, we have two of particular importance:

– Pilgrims of Saint Antoninus Martyr of Piacenza (570 AD)
– Nonnos of Panopolis (440 AD)

The first, from the pilgrims of San Antonino Martyr de Piacenza, is from around 570[26] as we saw in a previous chapter. This reference tells us about the presence of the Sudarium of Christ in the Holy Land at a time incompatible with the dating of the Sudarium of Oviedo. However, to be cautious, this reference tells us about the Sudarium of Christ and we do not have full evidence that it is the Sudarium of Oviedo.

But from the second one, which is an even older reference, we do have an indication that the Sudarium found today in Oviedo existed in the 5th century before the oldest limit of the C14 dating (642 AD). This is the paraphrase of the Gospel of John written by Nonnos of Panopolis. As we have seen in previous chapters, this author wrote his Paraphrase to St. John, XX 26 between 431 and 440 AD. He refers to the Sudarium of Christ and indicates that it had a knot. The mention by this writer of the "knot" of the Sudarium in the sepulchre of Christ and the observation of wrinkles also due to a knot in the Sudarium of

[25]Flori, J. (2004). *Guerra Santa, Yihad, Cruzada. Violencia y religión en el cristianismo y el islam*. Biblioteca de Bolsillo, n° 13, Universidad de Granada. pp. 129–130.

[26]Guscin, M. (2006). *La Historia del Sudario de Oviedo*. Ayuntamiento de Oviedo. Oviedo Avilés. p. 56.

Oviedo is an indication that, with high probability, it is the same cloth in both cases. The text of Nonnos of Panopolis is better understood when we know the Sudarium of Oviedo and the knot of this Sudarium is justified with the text of Nonnos. It should be borne in mind that the reference to the knot is unique and does not appear in the canonical gospels. It is not, therefore, a detail that could be included in a possible forgery of the relic. Moreover, the indication of a knot in the Sudarium of Oviedo was unnoticed and only appears in the course of a specific investigation when illuminated with grazing light or a laser line in search of old wrinkles as we saw in a previous chapter. Though if Nonnos at that time were describing a sudarium that he mistakenly considered to be that of Christ, at least he is describing a sudarium with a unique element that is also present in the Sudarium of Oviedo and that identifies it with the one that existed at that period. In that case, this is the strongest evidence that the Sudarium of Oviedo must be prior to the year 750 when it was dated, invalidating the measure.

6.5 Causes of Wrong C14 Dating

We now analyse here the possible causes of an erroneous dating result with the C14 method.

First of all, we must remember that the presence of C14 does not have its origin in the mineral world but in the biological world. Coal mines do not contain the C14 isotope. This is produced in the atmosphere by the collision of cosmic rays with nitrogen. It can also be produced by neutron bombardment on carbon 13 (C13) and: oxygen 17 (O17).

$$\text{Nitrogen } {}_0^1 n + {}_7^{14} N \rightarrow {}_6^{14} C + {}_1^1 p \tag{6.3}$$

$$\text{Carbon } {}_0^1 n + {}_6^{13} C \rightarrow {}_6^{14} C \tag{6.4}$$

$$\text{Oxygen } {}_0^1 n + {}_8^{17} O \rightarrow {}_6^{14} C + {}_2^4 \alpha \tag{6.5}$$

As we said above, plants breathe C14 and incorporate it into the food chain. In our case it is the same flax that incorporates it. When the organism dies, it stops incorporating new C14 and its content decays radioactively, reducing it by half every ~5730 years.

For a C14 dating to be reliable, it is necessary that the original proportion of the living organism (flax) is not altered after its death except by radioactive decay. Later contamination with biological material would alter the dating. The problem of contamination, defined as adhered foreign matter, is often invoked as the origin for a wrong dating. We analyse it in the following section.

6.5.1 Biological Contamination

A linen fabric is made up of threads, which are made up of fibres. Each thread can have hundreds of fibres that have a tubular structure. The wall of the fibres is mainly formed by cellulose. The outermost layer, PCW (Primary Cell Wall), has a thickness of about 0.2 microns and is made up of polysaccharides that are less chemically stable than the inner cellulose. In certain areas cellulose is in crystalline form and in others, it is amorphous. The most stable part is the crystalline area.

The cellulose is chemically a chain of monomers of formula $C_6H_{10}O_5$ (Fig. 6.10).

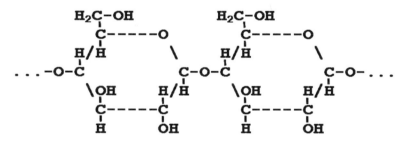

Figure 6.10 Chain of monomers of the cellulose.*

The monomer length is 5 Å = 5 × 10^{-4} μm = 5 × 10^{-7} mm and the chain can have 3000 to 4000 rings: 2 μm = 0.002 mm.

Now, let's assume that biological contamination is added; only added but not chemically bonded, to the original flax fibres.

The endless research carried out by Felipe Montero allowed him to discover some black fibres in the threads of the rest of the sample used for the C14 dating. Figure 6.11 allows realizing the proportion of black fibres on a thread. The original image can be found in Montero[27].

Figure 6.11 Scheme of the black fibres around a thread of the Sudarium.

He observed the surface of the black fibres under the Scanning Electron Microscope (SEM) and found that the coating was discontinuous.

Montero also analysed the black fibres with microanalysis, and he found the composition of the black fibres to have the same elements as the cellulose, i.e. mainly carbon and oxygen (Fig. 6.12). If we take the height of the oxygen peak as reference, we observe that the proportion of carbon is lightly larger in the black fibres than in the clean fibre. However, if we focus on the secondary peaks, we find presence of inorganic elements such as silicon, calcium, aluminium, iron and copper. This finding excludes the origin of the coating in the activity of microorganisms because they do not have metals in their composition. The unexpected finding was that the Spanish olive oil has traces of these metals.

[27]Montero, F. (2012). *The Date of the Linens for the Method of the C14. Particular Case of the Shroud [sic] of Oviedo. Proceeding of the I International Congress for the Shroud.* Valencia. 28–30 de April. available https://tienda.linteum.com/audiovisuales/99-i-con.html. Also, Montero F. (2015). 4th convention of the CES in Cordoba 2015. Video record available on https://tienda.linteum.com/audiovisuales/153-iv-convencion-2015-del-ces-los-lienzos-de-cristo.html.

He also compared the Raman spectrum of the clean fibres with the Raman spectrum of black fibres, and he found they are similar for both with some additional hills at 1330 cm^{-1} and 1590 cm^{-1} in the black fibres. These peaks are characteristic of soot, and he concluded that the coating is amorphous carbon.

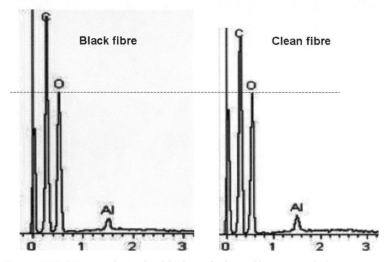

Figure 6.12 Microanalysis for black and clean fibres rescaled to oxygen height.

The β radioactive activity characteristic of C14 was also measured and it was higher for the black fibres. This proves that they had more C14 isotope than the original fibres.

All these data prove that there is amorphous carbon contamination in some fibres of the Sudarium. Montero proposed that the contamination is due to the soot produced by the poor combustion of the olive oil that lighted the Holly Chamber for centuries. In fact, the lighting of the Holy Chamber in the Cathedral of Oviedo used olive oil in the lamps at least since the 16th century until the arrival of electricity in the 20th century[28].

---

[28]Montero, F. (2012). *The Date of the Linens for the Method of the C14. Particular Case of the Shroud [sic] of Oviedo*. Proceeding of the I International Congress for the Shroud. Valencia. 28–30 de April. Also, Communication of F. Montero in the 4th convention of the CES in Cordoba 2015. Available on https://tienda.linteum.com/audiovisuales/153-iv-convencion-2015-del-ces-los-lienzos-de-cristo.html.

Now, let's evaluate, in a rough calculation, how much this contamination can shift the C14 dating. We consider that this soot or other type of substance is only deposited on the fibres, but it is not linked by chemical bonds.

Although contamination can shift the age of an archaeological sample, quantitative evaluation is often not performed. In order to calculate the shift, we remark that the hypothetical biological contamination could have lived and died during a certain period. If the contamination is modern, it would have more C14. If the contamination is old, it would have less C14. Moreover, the hypothetical biological contamination could have a percentage of carbon among their components or, even to be pure carbon (very improbable). In Table 6.7, we show the percentage of contamination, with respect to the original fibre weight, that a fibre of the first century must have to give a date of 750 AD. It depends on the living year of the contamination and the quantity of carbon in the chemical composition of the contaminant.

Table 6.7 Percentage of contamination

	% of Carbon in contamination			
Year of contamination	**100%**	**75%**	**44%**	**10%**
1950	23%	31%	53%	233%
1936	24%	31%	54%	236%
1800	27%	36%	61%	269%
1600	34%	45%	77%	337%
1400	45%	59%	101%	446%
1200	65%	87%	148%	652%
1000	119%	158%	270%	1189%
800	602%	802%	1368%	6017%

In Table 6.7, we show that contaminations of ancient time require a huge amount of foreign matter to change the date by seven centuries.

In the most favourable case, if it were modern pure carbon of biological origin, shifting the date about 750 years would require a proportion of 23% of contamination adhered to the fibres of the flax.

With this data let's evaluate the case of the contamination by olive oil. We need to determinate the year of the oil contamination. If we consider that the olive oil lamps were lighted since the 16th century to the 20th century, we can assume that the average time of the contamination could be 1800 AD. With this hypothetical case we check Table 6.7 along the row of 1800 year. The oil has a 77% of carbon. This percentage fall between columns 100% and 75% and it should be closer to the value of required contamination of 75%. We find that the contamination must be of about 35% of oil. However, Montero proposed that the contamination is not directly of olive oil but of the soot produced by the poor combustion of the oil. The smoke of the combustion of the oil is mainly carbon dioxide, carbon monoxide, hydrocarbons which incorporate nitrogen, and compounds also including metals. In general soot has an organic fraction with a carbon content of up to 60% and a high content of inorganic material[29]. Because the true composition of soot has a lower percentage of carbon than oil, the percentage of contamination must be increased for shifting the dating to the value required.

We can conclude that there is some contamination of the sample sent to the laboratory, but when Montero inspected the sample before the shipment, he found it mostly clean. A meticulous microscopic observation of the sample at 400× zoom before cutting it for shipment to the laboratory did not reveal any biological coating on the fibres of the threads, embeddings, strange threads, etc. In particular, there were traces of oil acids in practically zero quantities[30].

The laboratory also inspected the sample prior to testing and would have detected significant external contamination. Contamination of 20% or more can be excluded.

We note that the previous numbers correspond to the case that the cleaning processes carried out by the laboratories do not remove the contamination. If the cleaning process remove contamination, the shift of the date by added contamination will be even less.

---

[29]IARC, ed. (2010). *"Soot, as found in occupational exposure of chimney sweeps"* Chemical Agents and Related Occupations. IARC Monographs on the Evaluation of Carcinogenic Risks to Humans Volume 100F. p. 209.
[30]Montero, F. (2012). *The Date of the Linens for the Method of the C14. Particular Case of the Shroud [sic] of Oviedo.* Proceeding of the I International Congress for the Shroud. Valencia. 28–30 de April.

Our assessment on the adhered contamination is that it is unlikely to justify the dating error. It would be necessary that:

- The pollution is very modern.
- It must be rich in biological carbon.
- The decontamination process in the laboratory is not effective.

However, there is a possibility that contamination could resist the cleaning process much more than the original linen (explained later). In this case:

- The decontamination process eliminates the original cellulose and preserves the contamination.

The dating commissioned by EDICES was done in Beta Analytic and the cleaning treatment was the most exhaustive: "acid/alkali/ acid/cellulose extraction" (Table 6.8).

Table 6.8 Cleaning process applied in Beta Analytic

Chemical product	Concentration	Temperature	Time
Hydrochloric Acid	HCl 0.1 N (0.038%)	80°C	30 min.
Sodium Hydroxide	NaOH 2%	80°C	2 hours
Hydrochloric Acid	HCl 0.1 N (0.038%)	80°C	30 min.
Sodium Chlorite	$NaClO_2$ pH=3	70°C	?

Hydrochloric acid eliminates carbonates. Sodium hydroxide eliminates organic secondary acids. Sodium chlorite eliminates all the components except the cellulose. In some cases, the final treatment uses sodium hypochlorite (NaClO) at 2.5% for a duration of 30 minutes. Finally, the sample is dried to 105°C for a duration of 12 h[31].

In experiments carried out personally with contamination by beeswax vapour in the Institute Torres Quevedo of the *Consejo Superior de Investigaciones Científicas* in Madrid, the contamination was visible to the naked eye by the yellowish colour of the sample (18% contamination by weight). Observed at the Scanning Electron Microscope, (SEM) the wax was easily seen around the fibres and the microanalysis provided a clear increase of the

---

[31]Montero, F. (2012). *The Date of the Linens for the Method of the C14. Particular Case of the Shroud [sic] of Oviedo*. Proceeding of the I International Congress for the Shroud. Valencia. 28–30 de April.

carbon percentage. The cleaning methods were applied (without Sodium Chlorite) and they were relatively effective. To the naked eye, the fabric was bleached. The SEM also showed the absence of strung balls and the microanalysis indicated that the surface coating of the fibres had been removed. Finally, the weight of the samples after decontamination returned to the original value. All this indicated that the cleaning process of the radiocarbon laboratories was effective against the simply added contamination.

Montero on his own initiative applied the cleaning process used by the laboratories to linen of the 15th century and to modern linen. He verified that in the case of the old linen the weight of the sample was reduced much more (49.8%) than in the case of the modern linen (10%). By the way, we note that in the case of the sample for dating the Sudarium, the loss of weight communicated by Beta Analytic was of 58.4%.

In addition, Montero applied the cleaning process of the laboratory to the black fibres and he found that it was not effective. He verified that the fibres did not dissolve and that they did not lose their black coating.

Then, we are in the case that the decontamination eliminates the original cellulose and preserves the contamination. As we just said above, the older is the linen, the larger is the loss of matter. If the black fibres do not dissolve in the same degree as the clean fibres, the residue after cleaning that was introduced in the C14 detector had the modern black fibres almost intact and the original fibres reduced in more than a half. This would have increased the percentage of modern carbon and therefore would have rejuvenated the date of the sample. This contamination could have had a strong impact in the result of the dating.

To corroborate the results of Montero, we personally carried out experiments contaminating the samples with pure graphite. We used black chalks that the manufacturer describes as pressed carbon mixed with soot. We succeeded to contaminate up to a 5.3% in weight. We applied the full cleaning process, and the result is shown in Fig. 6.13. It is evident to the naked eye that the cleaning method used by the laboratory does not succeed in eliminating the graphite contamination.

Figure 6.13 Modern linen clean (left), contaminated with graphite (middle) and after decontamination process (right).

By the way, we confirm that modern linen only reduces its weight about 10% after the three first treatments: acid/alkali/acid. In this case we also used sodium chlorite as the last treatment, and we verify that it does not attack modern linen.

Now we will try to quantify the impact of this detected soot contamination. If we look at Fig. 6.11 and Fig. 6.14, we can estimate the area of black fibres versus the area of clean fibres. A coarse percentage could be 30% of black zone and a 70% of clean zone.

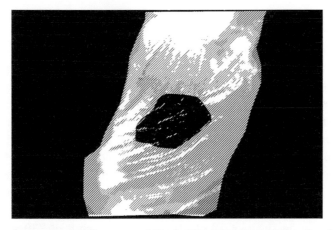

Figure 6.14 Scheme of a spot of black fibres in the middle of a thread found by Montero.

We assume that the cleaning attack reduces the original cellulose by 58.4%. Only 42% of the original cellulose survives the cleaning process. However, the black matter remains intact.

This implies that among the residue that reaches the C14 detector there is about 51% of black fibres and about 49% of clean fibres. But not all the bulk of the black fibre is contamination. The contamination is a coating of the original fibre.

As another easy estimation, we assume that the soot coating of the black fibres can be an extra 20% of the diameter of the fibre. The model is presented in Fig. 6.15. With this model, the percentage of the weight of the contamination versus the original material in a black fibre is near 50% (after deducting the central tunnel). We assume that the coating of the black fibre prevents the cleaning attack to the inner cellulose.

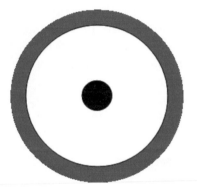

Figure 6.15 Model for the 20% of soot coating of the black fibres.

Now we can estimate how much percentage of mass contamination there is in the residue after the cleaning process. 51% are black fibres and 50% of each fibre is soot contamination. The original cellulose remains inside the black fibres and in the residue of the clean fibres after cleaning. Then, there is 34% of weight of soot on the original weight of the linen in the matter that is analyzed by the C14 detector. If we interpolate in Table 6.7, we obtain Table 6.9 where 34% of contamination living in the average year of 1850 AD can justify a shift of 750 years if the carbon content in the soot was 75%.

We see that the soot contamination detected present in the fibres of the Sudarium can justify the error in its C14 dating.

However, we realize that the hypothesis and assumptions can be discussed. The percentage of black fibres may have been overestimated. The thickness of the soot coating may have been overestimated too. And the carbon content in the soot may be less than 75% (it was estimated in 60%). With a carbon content of 60%, all contamination should occur in the future!

Table 6.9 Percentage of contamination

Year of contamination	% of carbon in contamination		
	75%	**60%**	**44%**
1950	31%	40%	53%
1936	31%	40%	54%
1850	34%	44%	58%
1800	36%	46%	61%
1600	45%	58%	77%
1400	59%	76%	101%

In fact, the weakest hypothesis is that of 30% of black fibres. Contamination of these fibres is only partial along the length of the fibre, the rest remaining clean. Although we used the images obtained by Montero to estimate the percentage of black fibres, we find incoherence with the percentage of graphite contamination carried out by us. With only 5.3% in weight our sample looks almost entirely black (Fig. 6.13) With 30% of soot, the Sudarium of Oviedo should look entirely black, and this is not the case.

A soot contamination of 5% only shifts the date about hundred years and a soot contamination not detected by the naked eye could only shift the date some decades. However, this detected contamination can, in any case, contribute lightly to shift the true age of the Sudarium when it is obtained through the C14 method.

To obtain greater shifts, there are other possible processes that can change the original content of the C14 isotope in the composition of the Sudarium.

6.5.2   Chemical Modification

In addition to contamination as adhered foreign matter, the dating can also be altered by chemical modification. Carboxylation can be a type of chemical contamination. Carboxylation is the chemical bond of a molecule of CO_2 to the unscreened OH groups of the cellulose molecules. In each monomer ring there are three OH groups and each of them can trap a CO_2 group. In this type of contamination, the foreign matter is added to the original matter. But in this case, the chemical bond prevents the cleaning of the sample by laboratory cleaning standards.

Carboxylation was detected by Kouznetsov[32] and his Russian team when they simulated the Shroud of Turin fire of 1532. The presence of carboxyl groups and aldehydes were detected in the Shroud[33]. The reaction can be stimulated by the increase of temperature, but it can happen at low rate at room temperature.

The calculations yield that a full carboxylation of the three unscreened OH groups of all the monomers of the cellulose with modern atmospheric CO_2 is needed to shift the dating in 723 years. It would seem an explanation of the case of the Sudarium. However, full carboxylation implies an increase of weight of the fabric by 81%[34] and it would be detectable by spectral analysis. The hypothetical carboxylation should happen along the years when the Sudarium was in contact with the surrounding air and not only in the 20th century. If this effect happens at room temperature, every archaeological fabric that was not in a sealed container, but that has been in contact with the surrounding air should give a shift in its dating. This problem does not seem to occur. On the other hand, the findings of Kouznetsov et al. had a reply[35]. We also performed a simple test on a sample of modern flax by heating it to 200 Celsius for 10 minutes in a natural atmosphere and found no weight gain after baking.

There is another possible chemical modification. A real structural change that hardly increases the weight of the fabric is

---

[32]Kouznetsov, D., et al. (1996). *Effects of fires and biofractionation of carbon isotopes on results of radiocarbon dating of old textiles: The Shroud of Turin.* Journal of Archaeological Science. 23. pp. 109–121.

[33]Schwalbe, L. A., and Rogers, R. N. (1982). *Physics and Chemistry of the Shroud of Turin, A Summary of the 1978 Investigation*, Analytica Chimica Acta, Vol. 135, February 1982, pp. 3–49.

[34]Salet, G. (1994). Supplement to the Lettre Mensuelle of the CIELT. n.53. May. 1994. p. 4.

[35]Jull, A. J. T. (1996). *A Comment on "Effects of Fires and Biofractionation of Carbon Isotopes on Results of Radiocarbon Dating of Old Textiles: The Shroud of Turin",* by D. A. Kouznetsov et al. Journal of Archaeological Science. 23. pp. 157–160.

the Isotopic Exchange: That is the basis of Francisco Alconchel's theory[36]: atmospheric carbon passes to cellulose and vice versa. Although he refers to the dating of the Shroud, the process can be applied to the Sudarium too. He proposes carboxylation only as an intermediate step. According to Alconchel, the incorporated CO_2 delivers its C14 to the cellulose monomer and a C12 of the cellulose monomer passes into the incorporated CO_2 and subsequently the CO_2 is released and returns to the atmosphere, having left its C14 in the tissue. Thus, the proportion of C14 would increase while maintaining the chemical formula and would be indistinguishable for the dating system. He proposes the reaction:

$$^{14}CO_2 + {}^{12}C_6H_{10}O_5 \Leftrightarrow {}^{12}CO_2 + {}^{14}C^{12}C_5H_{10}O_5 \tag{6.6}$$

In which one C14 isotope is exchanged by one C12 isotope. This approach is difficult to verify, but if it happens at room temperature, we can raise the same objection as for the carboxylation: every archaeological fabric that has been in contact with the surrounding air should give a shift in its dating.

According to the theoretical calculation of Alconchel, the isotopic exchange increases with the temperature because it would be endothermic. As we saw above, Moroni's team heated a sample of the Sudarium to verify the effect and the result was a shift to be older and not to be younger. This hypothesis needs more research.

6.5.3 Fungi and Other Biological Activity

Under suitable conditions micro-organisms which inhabit soil, water, and air can develop and proliferate on textile materials. These organisms include species of fungi, bacteria, and algae. Textiles made from natural fibres are generally susceptible to bio-deterioration[37]. Some bacteria, fungi, mould and yeasts are

---

[36]Alconchel, F. (2012). *Isotopic exchange between the Chambery fire smoke and the Shroud of Turin.* Proceeding of the first International Congress for the Shroud. Valencia 28–30 de April 2012.

[37]Hamlyn, P. F. (1990). *Why micro-organisms attack textiles and what can be done to prevent this happening.* Published in the Dec 1998 issue of North West Fungus Group (NWFG) Newsletter. This article originally appeared in Textiles, 1990, 19, pp. 46–50. http://fungus.org.uk/nwfg/rot.htm.

capable of metabolizing carbon dioxide and carbon monoxide as their sole source of carbon[38]. Bacteria fix CO_2 like plants (Calvin-Benson cycle) or through the reductive carboxylic acid cycle (Arnon cycle, anabolic). If this carbon is absorbed into the cloth, the content of C14 would be distorted.

We saw in the paragraph dedicated to DNA that the analysis discovered traces of a fungi *Arachnomyces minimus* that grow on rotten wood.

The research carried out by Felipe Montero was also focused on the possible presence of fungi and bacteria. It allowed him to discover a fungus in fibres of the Sudarium[39]. Montero also detected some staphylococcus. He requested DNA identifications.

He succeeded to grow the fungi in the black fibres and could identify them. It was *Cladosporium* which is a genus of fungi including some of the most common indoor and outdoor moulds. Many species of *Cladosporium* are commonly found on living and dead plant material (as linen). *Cladosporium* spores are wind-dispersed, and they are often extremely abundant in the outdoor air. Indoors *Cladosporium* species may grow on surfaces when moisture is present[40]. A particular study in the Library of the cathedral of Jaén (Spain) shows that the species *cladosporioides* is the main species detected (51'5%) in the records during the three months of winter[41]. This species can grow under low water conditions and at very low temperatures[42]. The study is highlighted because it was developed in a cathedral where there were very old material present, sometimes in an advanced state of deterioration. The conditions can be close to those of the cathedral of Oviedo.

Although we must not forget that the number of micro-biological genres that do not grow in cultures is most numerous, other analyses show that there is no living mycological activity

---

[38]Moon, P. (2013). *Coloured Dissolved Organic Matter (CDOM) contamination, mould damage, biocides and the carbon-14 dating of the Shroud of Turin*. Little Aston,

[39]Montero, F. (2015). 4th convention of the CES in Cordoba. Video record available on https://tienda.linteum.com/audiovisuales/153-iv-convencion-2015-del-ces-los-lienzos-de-cristo.html.

[40]https://en.wikipedia.org/wiki/Cladosporium.

[41]Sánchez Escabias, V. (2016). *Estudio aerobiológico de las esporas fúngicas en interiores, influencia sobre la salud y procesos de biodeterioro*. Jaen University. Faculty of Experimental Sciences.

[42]https://en.wikipedia.org/wiki/Cladosporium_cladosporioides.

on the Sudarium. In 1990 and 1994 the EDICES member Enrique Monte[43] took 32 samples and he processed them in micro-biological culture at 25°C and in dark and light conditions for two months[44]. Only one of the 32 samples showed growth of *Penicillium* fungi. The conclusion was that there is low risk of biodegradation of the Sudarium of Oviedo. Following the detection of the black fibres, microbiological analyses were also undertaken to culture the possible microorganisms in the black fibres. They tried with the traditional standardized cultures for bacteria (TSA) at 30°C and for fungi (PDA) at 28°C. The results were negative growths after 60 days of incubation. This result indicates that microorganisms have not grown because they are dead or because they are not viable in the means or conditions used. As Montero reports, the result of genetic analysis did not find remains of bacterial DNA, at least of a size bigger than 800 pb that was the limit of the experimental procedure. Only a dermatophyte *Malassezia furfur* grew which is responsible of seborrhea dermatitis in humans and must have come from contamination of human origin.

The mycological elements found in the Sudarium of Oviedo along the decades of research are:

- *Arachnomyces minimus*
- *Cladosporium (cladosporioides?)*
- *Penicillium (Brevicompactum?)*
- *Malassezia furfur*

The obvious presence of these microorganisms requires specific conservation conditions to prevent its appearance. To stress on the gravity of this risk, we should mention the anecdotal event of a test carried out without any scientific rigour on a modern piece of linen contaminated with the mould of ham! We put together the fabric and the mould in a plastic box and let the time pass. After about a year, we did not notice an evident change. But when we inspected the box the following year, we observed the total destruction of the linen fabric that was reduced to dust (Fig. 6.16).

---

[43]Professor of Microbiology and Genetic. University of Salamanca. Spain.
[44]Monte, E. (1994). *Estudio Micológico del Sudario de Oviedo*, Proceeding of the 1st International Congress of the Sudarium of Oviedo. pp. 91–93.

Figure 6.16 Biodegradation of modern linen by ham mould.

Fortunately, the quantity and activity of the microorganisms in the Sudarium of Oviedo is almost imperceptible and the biodegradation risk of the fabric is not high. Thanks to the new storage system this risk has been further reduced.

In any case, its contribution to an error in the C14 dating is very small. We recall that the detected presence of metal residues in the composition of black fibres excludes their origin in the microorganisms.

6.5.4 Irradiation by Neutrons

The result obtained in the dating of the Sudarium implies that there is approximately 7.2% more C14 in the linen used for weaving it than would be expected if it were from the time of Christ. Current physics knows the mechanism of production of this isotope based on the bombardment of neutrons on biological matter containing nitrogen (N14), the isotope 13 of carbon (C13) and the isotope 17 of the oxygen (O17) (see Equations 6.3, 6.4 and 6.5). These three elements are in the linen of the Sudarium:

C13 and O17 as part of the specific structure of the cellulose and N14 is present around 0.1% due to impurities. Although nitrogen is not an element of pure cellulose, it is notably more effective in capturing neutrons. While the impurities of nitrogen should have a different chemical behaviour, Arthur Lind[45] verified that after neutron irradiation the new C14 generated remains integrated in the cellulosic structure and is not removed by cleaning methods in the laboratories.

Therefore, if the fabric of the Sudarium were to undergo neutron irradiation, it would increase its C14 content and would give a younger date. The immediate objection raised was about the origin of the supposed neutron source. The author personally had this experience when he interviewed Dr Hans-Arno Synal, head of the Zurich dating laboratory in 2014, who participated in the measurement of the Shroud of Turin in 1988 (Fig. 6.17).

Figure 6.17 The author with Dr Hans Arno in the Swiss Federal Institute of Technology in Zurich.

When asked about the consequences of the neutron bombardment, Dr Hans Arno immediately wondered where the neutrons would come from. But as soon as he was told that the question was whether it would be possible to detect any effect of a neutron irradiation other than the increase in C14, he

[45]Lind, A. C., et al. (2010). *Production of Radiocarbon by Neutron Radiation on Linen.* Proceedings of the International Workshop on Scientific Approach to the Acheiropietos Images. ENEA Frascati.

concentrated on finding other isotopes present in the linen that would transmute and at the same time be sufficiently stable. Our approach was as follows.

- How much neutron flux does it take to change the dating of a linen of the year 33 AD to the year 750 AD?
- Is it possible to detect this irradiation by an effect other than C14?

These questions are typical of physics but not of metaphysics. If these questions have a positive answer and the result of the tests is that there was no irradiation, the topic ends. If the result of the test is that there was neutron irradiation, then the next question to ask is:

- Where did the neutron flux come from and when could it have happened?

The hypothesis of a neutron irradiation arose immediately after the Shroud of Turin dating. Then, the neutron flux was calculated for the case of the Shroud. In addition, years after, Robert Rucker went one step further. He is an experienced nuclear engineer who worked for 38 years on nuclear reactors at General Atomics in San Diego and for the DOE (Department of Energy). Rucker used the MCNP (Monte Carlo Neutron Particle) program developed by the Los Alamos National Laboratory to perform neutron flux analysis. He simulated the case of an emission of neutrons from the body wrapped in the Shroud in a limestone tomb and tuned its intensity to explain the case of the Shroud sample very accurately[46]. The model predicted an important gradient of C14 content along the Shroud (Fig. 6.18). If a sample of the middle of the Shroud were dated by C14, it would give a very modern date, even into the future. Rucker's results explain the C14 gradient in the three samples taken from a corner of the Shroud that were carbon dated. In addition, he calculated how much neutron intensity the Sudarium would receive if it were about a metre from the body (Fig. 6.19).

---

[46]Rucker, R. (2014). *A. MCNP Analysis of Neutrons Released from Jesus' Body in the Resurrection.* Proceedings of the Shroud of Turin Conference. St Louis. October 9–12, 2014. www.stlouisshroudconference.com. Rucker, R. (2020). *Understanding the 1988 Carbon Dating of the Shroud.* Paper 25, www.shroudresearch.net/research.

The neutron flux received by the Sudarium in this model would also explain the shift of dating for the cloth of Oviedo. This is a very suggestive model because it explains concurrently the shift of dating for the two cases: The Shroud and the Sudarium.

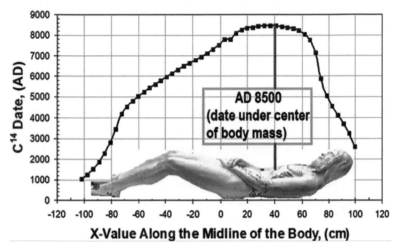

Figure 6.18 Date corresponding to the C14 content along the Shroud of Turin if a neutron irradiation comes from the body (© Rucker).

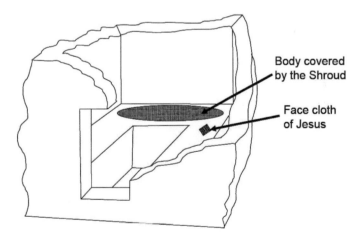

Figure 6.19 Shroud and Sudarium receive the neutron irradiation that justifies the dating error (© Rucker).

To estimate the required neutron flux to justify the dating of the Shroud, it is necessary to identify the reactions that produce new C14. Thomas J. Phillips[47] proposed a dose of 2×10^{16} neutrons cm^{-2} for shifting the date of the Shroud of Turin from the year 33 AD to the year 1350 AD. However, according to Robert E.M. Hedges[48], the dose proposed by Phillips is much too high, as he has not included the neutron captured by nitrogen in the cloth. If we consider the transmutation of nitrogen, a thermal neutron flux of 2×10^{13} cm^{-2} (that is, 1,000 times less) would be enough. If we add also the contribution of O17, the flux decreases still more ($\approx 1.94 \times 10^{13}$ cm^{-2}). Thus, we wonder if this weak flux, necessary to prove the neutron irradiation, is enough to produce other isotopes in a detectable quantity.

Phillips calculated the dose assuming that the new C14 came from the capture of a neutron by the C13 isotope and proposed several neutron reactions that could have altered the content of other isotopes in addition to C14[49]. He identified chlorine 36 and calcium 41 as possible indicators of such irradiation as indicated in Table 6.10.

Table 6.10 Properties of parent/daughter isotopes

Parent isotope	Natural abundance (%)	Neutron capture cross section (barns)	Daughter isotope	Daughter isotope Half-life (years)	Foreseen parent /daughter ratio
^{13}C	1.11	0.0009	^{14}C	5730	1.2×10^{-10}
^{35}Cl	75.77	43	^{36}Cl	3.0×10^5	8×10^{-7}
^{40}Ca	96.94	0.40	^{41}Ca	1.0×10^5	8×10^{-9}

As we said, the predicted parent/daughter ratio proposed by Phillips and indicated in Table 6.10 is calculated for a flux of 2×10^{16} neutrons cm^{-2} irradiating a new linen cloth. If Hedges is correct, the ratio of new isotopes would be lower. Lind, et al.[50] found experimentally that a neutron dose of

---

[47]Thomas J. Phillips of the High Energy Laboratory of the American University of Harvard.

[48]Hedges, R. E. M. (1989). *Shroud irradiated with neutrons? Hedges Replies.* Nature Vol. 337. n° 6208. 16 February 1989. p. 594.

[49]Phillips, T. J. (1989). *Shroud irradiated with neutrons?* Nature Vol. 337. n° 6208. 16 February 1989. p. 594.

[50]Lind, A. (2010). *Production of radiocarbon by neutron radiation on linen,* Proceedings of the International Workshop on the scientific approach to

1.07×10^{14} n/cm^2 could produce a quantity of C14 that could change the C14 date of the Shroud from 33 to the interval 1260 to 1390 AD. It is a dose about 5 times higher than the estimation of Hedges.

In any case, following the Phillips – Hedges discussion, the unstable isotopes most likely to be found in the Shroud of Turin are Chlorine-36 (Cl36) and Calcium-41 (Ca41). Cl36 is almost nonexistent in the environment. There is an atom of Cl36 among 10^{12} atoms (1000 billions) of stable isotopes of chlorine. However, Cl36 can be generated by neutron irradiation of the abundant chlorine-35 (Cl35) and its half-life is about 300 000 years. Therefore, once created, Cl36 remains for many thousands of years. Ca41 is almost nonexistent in the environment. There is an atom of Ca41 among 1.25×10^{14} atoms (100000 billions) of stable isotopes of calcium. However, it can be generated by neutron irradiation of the common calcium-40 (Ca40) and its half-life is about 100 000 years. Therefore, once created, Ca41 also remains for many thousands of years.

Although chlorine and calcium are not present in pure cellulose, they can be impurities in linen. According to Thomas McAvoy[51], after washing a modern piece of linen, its chlorine content was measured as 56 ppm (parts per million). This chlorine is assumed to be present in organic contamination. He verified that this modern linen, once irradiated, contained the Cl36 isotope at a level measurable through mass spectroscopy. With a dose of 1.07×10^{14} n/cm^2 (that of Lind), the Cl36/Cl ratio in the modern linen that was irradiated was over 2000 times the upper limit estimated for naturally occurring Cl36 in linen. In the case of the Sudarium the expected ratio should be lower because the assumed dose should be lower than in the case of the Shroud. Cl36 can still be measured, and the measurement should still give a ratio significantly above background (if there has been no loss or gain of Cl along its history). Table 6.11 shows the predicted and experimental Cl36 ratio for different neutron doses.

the Acheiropoietos Images, ENEA Frascati, Italy, 4–6 May 2010. http://www.acheiropoietos.info/proceedings/LindWeb.pdf.

[51]Thomas McAvoy. Professor Emeritus, Institute for Systems Research, Department of Chemical and Biomolecular Engineering, and Bioengineering Department, University of Maryland, College Park MD, 20742, USA. Personal communication 05/10/20.

Table 6.11 Cl36 parent/daughter rate

Author (cloth)	Dose n cm^{-2}	Daughter/parent ratio
Philips (Shroud)	2×10^{16}	8.00×10^{-7}
Lind/McAvoy (Shroud)	1.07×10^{14}	2.61×10^{-9} (experiment)
(Sudarium)	9.80×10^{13}	1.10×10^{-9}
Natural ratio	—	$\leq 10^{-12}$

The drawback is, however, that to measure whether there is chlorine or any of its isotopes in the Shroud or in the Sudarium, we need to "burn" a quantity of fabric similar to that used for the C14 dating (>10 mg). In fact, the sample used for the Cl36 detection experiment was an order of magnitude larger than in the case of the Shroud (151 mg).

Was there another non-destructive test that could be used to discover if neutron irradiation occurred? Well, in the case of the Shroud, the neutron irradiation, according to the model of Rucker, would not have been uniform, because of the large differences between the location of various parts of the body and the parts of the Shroud. These differences result in large gradients in neutron dose received by various parts of the Shroud, resulting in differences in created C14. In the case of the Sudarium, the gradient would be lower than in the case of the Shroud. We saw the large gradient for different parts of the Shroud up to seven times larger in the middle than in the edge (Fig. 6.18). In the case of the Sudarium, it would have been about a metre from the body, and the relative difference between opposite edges of fabric would be less than the double. The gradient results in a significant difference in C14 content in the case of the Shroud and it would support dating differences for the Italian cloth but, in the case of the Sudarium of Oviedo the gradient would be small and its C14 dating should be more uniform.

Finally, Maurizio Bettinelli and his team found a change in the FTIR (Fourier-transform infrared) spectrum of irradiated linen[52]. They analysed irradiated ancient linen provided by the

---

[52]Bettinelli, M., et al. (2000). *Impiego di Tecniche Chimico–Fisiche per lo Studio dell'Invecchiamento delle Fibre di Lino*. Worldwide Congress Sindone 2000. Orvieto, 27–29 August 2000.

Lyon Textile Museum. It was a fabric of a Lyma mummy which had been previously dated by C14 to be 160 ± 60 BC by the Radiocarbon Dating Centre of the C. Bernardi University of Lyon. The piece of fabric was irradiated in the nuclear reactor of the University Luis Pasteur in Strasbourg with an integrated flux of 1.16×10^{13} n/cm^2. We note that the dose in this case was an order of magnitude lower than in the test to find the Cl36 isotope.

The FTIR spectrum in the region between 1500–1750 cm^{-1} for the mummy linen can be compared before and after irradiation. The peak at 1635–1650 cm^{-1} is characteristic of the carbonyl group (C=O) and the band at 1705–1715 cm^1 is characteristic of the carboxyl group (–C=OOH). They slightly increase following neutron irradiation (Fig. 6.20).

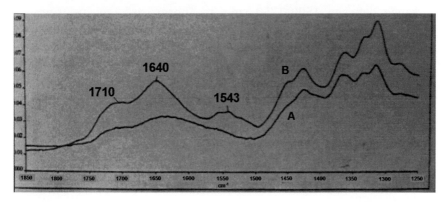

Figure 6.20 FTIR spectrum of mummy linen. Sample A original and Sample B irradiated with neutrons. Image processed from Bettinelli (2000).

In our opinion, however, the more significant change is the appearance of a peak at 1543 cm^{-1} that the authors of the experiments do not comment on. This infrared wave number is not associated with a vibration mode of a molecular group of the cellulose. A decrease in frequency with an increase in mass of the elements implicated in the vibration could explain the vibration at 1543 cm^{-1} if the C=C bond involves two carbon atoms of the C13 heavier isotope (C13=C13). In this case, the frequency of the carbonyl group would be reduced to 1543 cm^{-1}. The capture of neutrons by the most abundant C12 isotope provides some quantity of new C13. These new C13 atoms could bond to each other. However, the bonding of two C13 atoms would be much less probable than the bonding of a C12 with a C13 (C12=C13) because of the greater quantity of C12. The associated frequency of a C12=C13 bond is reduced to around 1610 cm^{-1}, by the increase in mass of the C13 component. Therefore, we should observe a large peak at this frequency if the production of C13 were the origin of the vibration at 1543 cm^{-1}—but it is not observed. In fact, with the specified neutron flux, the new bindings C12=C13 and C13=C13 would be almost undetectable. In short, we do not have a clear explanation for the 1543 cm^{-1} peak.

As a last analysis, we equalize levels at both sides of the range between 1500–1750 cm^{-1} to "normalize" the spectrums and we compare them for the mummy sample before and after irradiation and also for the Sudarium. The spectrum of the Sudarium shows an increase at the 1640 cm^{-1} peak wave number with a height that is similar to the peak of the irradiated mummy sample. However, the other two peaks at 1705–1715 cm^{-1} and at 1543 cm^{-1} are not present (Fig. 6.21). The apparent increase in intensity at the 1640 cm^{-1} wave number for the Sudarium is not enough to draw any conclusion on whether or not it was exposed to neutron radiation. More research is needed to verify whether these frequencies are truly present and linked to neutron irradiation.

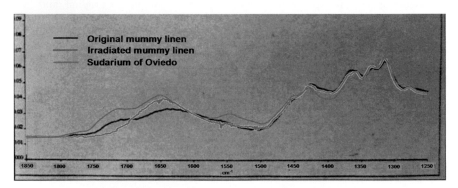

Figure 6.21 "Normalized" FTIR spectrum of a mummy linen. Sample original and sample irradiated with neutrons and sample of the Sudarium of Oviedo.

6.6   Other Dating Methods

There are few alternatives to C14 method for absolute dating of a textile. We can explore other possibilities to obtain indirect or relative dates, to help us know the true age of the Sudarium.

Sometimes, the analysis of the embroidery or prints of a fabric or the dyes used allow us to locate the time when the fabric was made. However, in the case of the Sudarium it is difficult to conclude anything about such a simple cloth. The signs of the torture suffered by the "victim" over whom it was placed are the only relevant aspect it has. The latter information is particularly important and significant, as discussed in previous paragraphs.

Moreover, we can analyse whether the type of clothing and the technology used to manufacture it could tell us something about the date. In 1997, Sophie Desrosiers, successor of the famous French expert Gabriel Vial, was consulted by Socorro Mantilla and Felipe Montero about the Sudarium of Oviedo. Different photos were sent to her for analysis, which allowed her to indicate that the Sudarium had probably been woven on a vertical loom with weights, but its date could not be specified because the use of this type of loom spans for centuries. She also added that the defects found in the Sudarium are similar to those found in Danish fabrics made on this type of loom between 400 BC and 500 AD. The vertical loom with weights is prior to the period of Romanization and is well known among the vestiges of the Roman era, but it is not exclusive to that time, and it does not determine a date. That is to say, the confection corresponds very well with the 1st century and Roman culture but does not exclude other dates and cultures. As we have seen in the chapter where the fabric was discussed, the Sudarium of Oviedo could have been woven in India as the Shroud of Turin. This tells us about the geographical origin but nothing about the time.

The age of the Sudarium of Oviedo has also been studied from another point of view involving the observation of its physical properties. On January 25, 2005, one month and 11 days before his death, the American chemist Raymond Rogers sent his report on the Sudarium to Father Carreira of EDICES.

Rogers was regarded as an authority on the study of the Turin Shroud and was a great expert on thermal effects. Around the same time his article was published in *Thermochimica Acta*[53] on the physicochemical differences between the dated sample of the Shroud and the rest of the cloth. Due to his capacity and prestige, some threads from the Sudarium were sent to him duly documented, to be analysed by him. He provided his report[54].

Rogers observed the sample from the Sudarium intensively with a petrographic (polarizing) microscope. He prepared the flax fibres from the Sudarium of Oviedo submerged in oil with an index of 1,515 and at 100X magnification (Fig. 6.22). He was able to perceive the small residual deposits of lignin in the knots of the flax fibres. There are dark rings around the fibres that are lignin not removed during bleaching. Moreover, and according to Rogers, ancient linen fibres that are 1300 years old or older, do not show traces of vanillin. The fibres of the Sudarium do not show traces of vanillin. Therefore, it should be 1300-year-old or older. But in the case of the Sudarium, the C14 result does not collide with the range of dates given by the alternative method of lacking vanillin. Considering this lack, the Sudarium could well be 1300 or 1400 years old, which would give a date of 600 or 700 AD, still agreeing with the C14 dating result of the Sudarium.

On the other hand, the observation indicated the use of a mild bleaching method used during Roman times, as described by Pliny the Elder. The technology is certainly different from the method more commonly used in modern or medieval times in Europe or for millennia in Egypt, where Natron (mixture of sodium carbonate and sodium bicarbonate) is used.

With the microscope, Rogers also observed the damaged or irradiated parts of the flax fibres since they manifest birefringence in contrast to the intact fibres (Fig. 6.23). Under cross polarized microscope, new flax fibres do not allow light to pass through along the segments between growth nodes and they appear perfectly black. On the contrary, irradiated fibres

---

[53]Roger, R. (2005). *Studies on the radiocarbon sample from the shroud of Turin.* Thermochimica Acta. 425. pp. 189–194.

[54]Roger, R. (2005). *The Sudarium of Oviedo: A Study of Fiber Structures.* Available in https://www.shroud.com/pdfs/rogers9.pdf. (October 2020).

with defects show zones of birefringence as we see in the flax fibres of the Sudarium of Oviedo. Rogers concluded that the fibres of the Sudarium show many defects caused by different kinds of radiation. For the observation, the fibres are immersed in oil of index 1,515 and at 800X magnification through crossed polarisers.

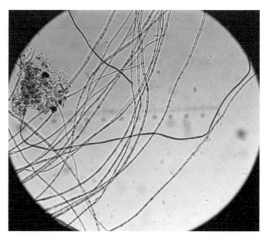

Figure 6.22 Flax fibres from the Sudarium of Oviedo observed by R. Rogers.

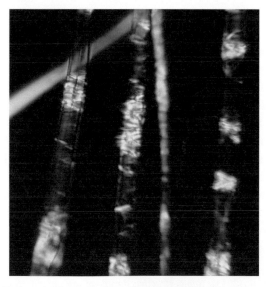

Figure 6.23 Defects in the Sudarium fibers (© Rogers).

The natural ambient radiation that comes from cosmic rays or the radioactive decay of terrestrial minerals produces defects by ionization in the crystalline structure of cellulose. When the cellulose of flax is formed in the vegetable plant, it hardly has these types of defects, but the exposure to natural radioactivity over the years accumulates these characteristic type of defects, increasing their number with the age of the fabric. Rogers found that the fibres of the Sudarium showed many defects caused by different types of radiation. The defects seen indicate significant age. Rogers compared this feature of the Sudarium to fibres of three fabrics:

- A fabric of 2000 years before Christ
- A fabric of 1620 after Christ
- The Shroud of Turin

The defects were similar to those of the Shroud of Turin in type and quantity. His conclusion is that the Sudarium of Oviedo shows all the physical and chemical properties of a very old linen sample. The types of defects observed in its cellulose crystals caused by ambient radiation are very similar to those of other ancient linens of known age. The fibres of the Sudarium show the same characteristics as those of the Shroud. He concludes that there is a non-negligible probability that the Sudarium is related in age and origin to the Shroud of Turin.

The indications of Raymond Rogers therefore place the Sudarium with some probability in the Roman period although they do not allow specifying a precise date.

Other alternative dating methods based on infrared (FTIR) and Raman spectroscopy has been developed by Giulio Fanti from the University of Padua[55]. He and his team have calibrated the variation of the infrared and Raman spectra of linen pieces according to their age. The precision of the method is considerably worse than that of C14 as it has an associated uncertainty of more than ±400 years compared to the ±30 years that laboratories associate with radio-dating.

---

[55]Fanti, G., et al. (2013). *Non-destructive dating of ancient flax textiles by means of vibrational spectroscopy.* Vibrational Spectroscopy. 67.

Another dating method also developed by Fanti[56] uses three mechanical parameters of the linen fibres. A single flax fibre 1–3 mm long is mounted on special supports and submitted to mechanical test consisting of multiple stress–strain cycles to determinate the breaking strength, the Young modulus and the loss factor. Once calibrated against some textiles of known age, this method estimates a date for the Shroud of 260 AD. The standard uncertainty of the method is about 270 years. The FTIR and Raman dating averaged with the mechanical result for the Shroud provide a date of 90 AD ±200 years[57]. It is well compatible with the death of Christ. Therefore, given the 700-year discrepancy between the expected time and that found by C14 for the Sudarium, it would be enough to confirm or deny what was found by the C14 method. This method has not been applied to samples of the Sudarium. If some of the already existing cut samples were available, this dating method could be applied because it is almost nondestructive.

Another alternative method that has been evoked for dating is "depolymerization" or the reduction of the degree of polymerization as a function of aging time. The linen is made of cellulose that is a polymer which tends to break down over time. The thermal, hydrolytic, photolytic, photochemical and oxidative processes increase the polymeric chain breaks, and it can be measured by wide-angle X-ray scattering (WAXS)[58]. The threads must not be older than about thirty centuries because of the saturation of the structural degradation with age. Its uncertainty for samples 20 centuries old is about ±100 years. However, this technique is still in the development phase and its main weakness is its dependence on unknown conservation conditions. A variation of a couple of Celsius degrees in the assumed conservation temperature implies about ±300 years of uncertainty. The main advantage is that this method can be performed nondestructively on a single thread of a few mm long because the tested area is about 0.2×0.5 mm^2. We can say again that this

---

[56]Fanti, G., Malfi, P. (2014). *Multi-parametric micro-mechanical dating of single fibers coming from ancient flax textiles.* Textile Research Journal. 84(7). pp. 714–727.

[57]Fanti, G., et al. (2015). *Mechanical ond opto-chemical dating of the Turin Shroud.* MATEC Web of Conferences, 36.

[58]De Caro, L., et al. (2019). *X-Ray Dating of Ancient Linen Fabrics.* Heritage 2019, 2, pp. 2763–2783.

method has not been applied to samples of the Sudarium, but it could be applied on existing samples because it is nondestructive.

6.7   Conclusion on Dating

The Sudarium of Oviedo has been dated by the C14 method three times giving the year 748 AD as the most probable one. Its flax would have been harvested between the year 642 AD and the year 890 AD, if we take in account the 95% confidence for the wider range among the three dating tests.

On the other hand, the results of the C14 dating technique are rejected by the archaeologists in about 20% of cases. Therefore, and despite the C14 result of the Sudarium of Oviedo, the hypothesis that the Oviedo Sudarium is that of Christ can still be considered.

Other analyses point rather to an age that is perfectly compatible with the tradition that relates the cloth to the death of Christ. However, their low precision does not lead to a direct contradiction with C14 dating.

For example, the death suffered by the victim whose head was wrapped in the Sudarium of Oviedo must be a crucifixion. It is the execution that best fits the evidence presented on the cloth. The Roman crucifixion was abolished in the Christian West more than 300 years before the C14 dating.

The strongest evidence against the C14 dating is the reference of Nonnos of Panopolis from the 5th century. This writer describes a knot in the Sudarium of Christ. The presence of the same singularity in the Sudarium of Oviedo identifies it with the one Nonnos knew in his time. If this evidence is accepted, the Sudarium of Oviedo existed already in the 5th century, three centuries before the date of the C14.

Several causes for a wrong C14 dating have been proposed. There is a black fibres contamination which has been carefully analysed. It can come from soot of the incomplete combustion of olive oil used to illuminate the Holy Chamber for centuries. It can shift the date provided by the C14 method. Surely, it altered the true date of the Sudarium. However, a quantitative estimation of the shift associated to the detected contamination can hardly justify the whole deviation.

If carboxylation and isotopic exchange would have impact on C14 dating of textiles, the results of C14 dating for the whole cases of this type of specimens would be wrong and this is not the case.

The risk of contamination by microbiological organisms has been analysed. But it affects bio-deterioration rather than dating.

A common irradiation by neutrons in the sepulcher would explain the dating shift in the case of the Sudarium and in the case of the Shroud. It could be excluded or confirmed by detection of chlorine 36 isotope. But it requires burning tens of milligrams of fabric. Other effects of neutron irradiation demand more research.

The age of the linen of the Sudarium can be estimated by other experimental methods different of C14. The numbers of crystal defects and the content of vanillin were observed in the Sudarium by Raymond Rogers. He concluded that these features of the Sudarium are similar to those of the Shroud in type and quantity and must be of the Roman period.

Infrared spectroscopy (FTIR), Raman spectroscopy, mechanical parameters of single fibres and wide-angle X-ray scattering (WAXS) are some non-destructive methods that are being developed and could be applied to the Sudarium once matured.

6.8 Annex: Mathematical Estimation of the Contamination Impact

This section can be skipped by the reader not interested in mathematical demonstrations.

In the dating laboratory they measure the proportion m_{14} of C14 ($[^{14}C]$) with respect to C12 ($[^{12}C]$) in the sample to be dated:

$$\frac{[^{14}C]}{[^{12}C]} = m_{14}$$

And they assume a typical C14 concentration in the live flax. It is the initial concentration m_{014}:

$$m_{014} = \frac{[^{14}C]_0}{[^{12}C]} \text{living\_flax}$$

It is assumed that the evolution of concentration m_{14} follows the radioactive decay:

$$m_{14} = m_{014}e^{-\left(\frac{t}{8266}\right)} \Rightarrow \frac{m_{14}}{m_{014}} = e^{-\left(\frac{t}{8266}\right)}$$

And, with the measured concentration and the assumed initial one, the age t is obtained:

$$t = -8266\ln\left(\frac{m_{14}}{m_{014}}\right) = -8266\ln\frac{[^{14}C]}{[^{14}C]_0} \tag{A1}$$

The result must be corrected with the calibration curve obtained with dendrochronology (tree rings).

Taking contamination into account

Now let us assume that the measured amount of C14 is not of the original flax but a mixture of the original and contamination. We assume that the cleaning process does not alter the original carbon to carbon contamination ratio. In other words, the cleaning process acts on both compounds with the same efficiency. Therefore, the amount of C14 that reaches the detector is the sum of original and contamination:

$$[^{14}C] = [^{14}C]_{\text{original}} + [^{14}C]_{\text{contamination}} = [^{14}C]_l + [^{14}C]_c$$

The $[^{14}C]_{\text{original}}$ is the C14 in the original flax and it is supposed to have been decaying since the year 33. The ages are calculated with respect to 1950 by agreement. Therefore, the C14 rate of the original flax will be:

$$\frac{[^{14}C]_l}{[^{12}C]_l} = \frac{[^{14}C]_{0l}}{[^{12}C]_l}e^{-\left(\frac{1917}{8266}\right)} \Rightarrow [^{14}C]_l = [^{14}C]_{0l}e^{-\left(\frac{1917}{8266}\right)} \tag{A2}$$

We try to find the age shift as a function of the percentage of contamination. So, now we are going to introduce the ratio of carbon by mass to the total mass of the compound. In the laboratory, the sample is burned, and the resulting CO_2 is collected. This CO_2 resulting from combustion is then processed to obtain pure carbon. Finally; this pure carbon passes by the accelerator to count isotopes. Although everything that is not carbon in the sample is removed, we are interested in the total amount of contamination, carbon or not, because it is what we can observe directly in the sample before is processed. For the original part, which is flax (mainly cellulose $C_6H_{10}O_5$) the percentage of carbon is 44% of the total original flax weight M_l:

$$[^{12}C]_l = 0.44 \, M_l \tag{A3}$$

Similarly, for contamination we have:

$$\frac{[^{14}C]_c}{[^{12}C]_c} = \frac{[^{14}C]_{0C}}{[^{12}C]_c} e^{-\left(\frac{t_C}{8266}\right)} \Rightarrow [^{14}C]_c = [^{14}C]_{0C} e^{-\left(\frac{t_C}{8266}\right)} \tag{A4}$$

where now t_C is the age of the contamination compared to 1950. We call X the proportion of carbon by mass with respect to the mass of contamination (M_C = total contamination weight; X = percentage of carbon in contamination):

$$[^{12}C]_C = X \cdot M_C \tag{A5}$$

We know that the age t of the Sudarium has been 1200 years (=1950–750):

$$t = 1200 \text{ years}$$

And the laboratory calculated this age from the proportion of C14 found by the instrument with respect to the assumed original $[^{14}C]_0$. But actually carbon came from two sources, original flax $[^{14}C]_l$ and contamination $[^{14}C]_c$. The formula (A1) will now be:

$$t = -8266 \ln \frac{[^{14}C]_l + [^{14}C]_c}{[^{14}C]_0} \tag{A6}$$

The amount of C14 in living linen can be approximated by one atom of C14 for every trillion of C12.

$$[^{14}C]_0 = 10^{-12} \, [^{12}C] \tag{A7}$$

The amount of C12 in the sample to be dated will be according to (A2) and (A5):

$$[^{12}C] = [^{12}C]_I + [^{12}C]_C = 0.44 \cdot M_I + X \cdot M_C \tag{A8}$$

We replace in (A6) with (A7) and (A8):

$$t = -8266 \ln \frac{[^{14}C]_I + [^{14}C]_C}{10^{-12}(0.44 M_I + X.M_C)} \tag{A9}$$

And the numerator is replaced with their respective decay formulas (A2) and (A4):

$$t = -8266 \ln \frac{[^{14}C]_{0I} e^{-\left(\frac{1917}{8266}\right)} + [^{14}C]_{0C} e^{-\left(\frac{t_C}{8266}\right)}}{10^{-12}(0.44 M_I + X.M_C)} \tag{A10}$$

Considering approximation (A7) we would have:

$$[^{14}C]_{0I} = 10^{-12} \cdot [^{12}C]_I = 10^{-12} \cdot 0.44 \, M_I \tag{A11}$$

$$[^{14}C]_{0C} = 10^{-12} \cdot [^{12}C]_C = 10^{-12} \cdot X \cdot M_C \tag{A12}$$

When we substitute in (A10) with (A11) and (A12), all factors 10^{-12} are cancelled. This is like accepting that the proportion of C14 in live flax is the same as in the contamination. That is, the contamination is of biological origin and came from living organisms when it contaminated the original linen. The proportion of C14 in living biological organisms is measured against a standard by δ^{14} defined as:

$$\delta^{14} = \frac{[^{14}C]_I}{[^{14}C]_{ref}} - 1$$

But in practice, the quantity of the other C13 isotope is measured. It is stable and can be easily measured in any sample no matter how old it is. And it is assumed that the mechanism that can modify the proportion of the C14 isotope incorporated in biochemical processes with respect to that in the atmosphere is a mechanism that has the double impact on C14 than on C13 isotope. That is to say:

$$\delta^{14} = 2 \cdot \delta^{13}$$

Thus, it can be verified whether the original C14 proportion assumed for the biological organism to be dated corresponds to the actual one by measuring the proportion of C13 in the sample. Most organisms have a δ^{13} around –25 per thousand, and therefore a δ^{14}, around –50 per thousand:

The cancellation of the factor 10^{-12} is, therefore, a good approximation and the date is based on the age of the contamination and the percentage of carbon in the total contamination mass:

$$t = -8266 \ln \frac{0.44 M_l e^{-\left(\frac{1917}{8266}\right)} + X.M_c e^{-\left(\frac{t_c}{8266}\right)}}{(0.44 M_l + X.M_c)} \tag{A13}$$

Let's put the contamination mass M_C as a percentage of the original flax mass M_l:

$$\alpha = \frac{M_C}{M_l} = \frac{\text{mass of contamination}}{\text{mass de original linen}} \Rightarrow M_C = \alpha M_l \tag{A14}$$

Now we put the age t depending on the age of the contamination t_c and the percentage of contamination α:

$$t = -8266 \ln \frac{0.44 M_l e^{-\left(\frac{1917}{8266}\right)} + X.\alpha.M_l e^{-\left(\frac{t_c}{8266}\right)}}{(0.44 M_l + X.\alpha.M_l)}$$

$$= -8266 \ln \frac{0.44.e^{-\left(\frac{1917}{8266}\right)} + X.\alpha.e^{-\left(\frac{t_c}{8266}\right)}}{(0.44 + X.\alpha)} \tag{A15}$$

Since $t = 1200$ years, we can solve for α as a function of the age of contamination t_C:

$$1200 = -8266\ln\frac{0.44.e^{-\left(\frac{1917}{8266}\right)} + X.\alpha.e^{-\left(\frac{t_C}{8266}\right)}}{(0.44 + X.\alpha)}$$

$$\Rightarrow e^{-\left(\frac{1200}{8266}\right)} = \frac{0.44.e^{-\left(\frac{1917}{8266}\right)} + X.\alpha.e^{-\left(\frac{t_C}{8266}\right)}}{(0.44 + X.\alpha)} \Rightarrow$$

$$(0.44 + X.\alpha)e^{-\left(\frac{1200}{8266}\right)} = 0.44.e^{-\left(\frac{1917}{8266}\right)} + X.\alpha.e^{-\left(\frac{t_C}{8266}\right)}$$

$$\Rightarrow X.\alpha.e^{-\left(\frac{1200}{8266}\right)} - X.\alpha.e^{-\left(\frac{t_C}{8266}\right)} = 0.44(e^{-\left(\frac{1917}{8266}\right)} - e^{-\left(\frac{1200}{8266}\right)}) \tag{A16}$$

Then, for a age t_C of contamination we calculate the percentage of contamination α that provides the year 750 AD as the date of the contaminated Sudarium. Table 6.12 gives some values of α as a function of X and justifies Table 6.7.

Table 6.12 Percentage of contamination

Year of contamination	% of Carbon in contamination			
	100%	75%	44%	10%
1950	23%	31%	53%	233%
1936	24%	31%	54%	236%
1800	27%	36%	61%	269%
1600	34%	45%	77%	337%
1400	45%	59%	101%	446%
1200	65%	87%	148%	652%
1000	119%	158%	270%	1189%
800	602%	802%	1368%	6017%

Chapter 7

The Shroud and the Sudarium

International scientific research has found a series of decisive correspondences between the Sudarium of Oviedo and the Shroud of Turin. In this chapter we describe several features of the Sudarium of Oviedo that correspond to the features of the Shroud of Turin.

Many readers of this book know the Shroud of Turin well because the interest in the Sudarium often comes after the knowledge of the Shroud. There is a separate book in this series devoted to this cloth[1]. However, we must briefly describe the Italian sheet.

The Shroud of Turin is a 4.34 m long and 1.10 m wide old linen cloth, traditionally considered to be the burial sheet in which the body of Jesus Christ was wrapped and placed into the tomb. To determine whether this is true, the historical and scientific research done on the Shroud has been more than on any other ancient object. It carries a life-size unexplained image of visible discolouration of the front and back of a naked Man with the hands crossed over the abdomen. The dead Man has blood stains that are unmistakable signs of crucifixion and scourging.

---

[1]Fanti, G., and Malfi, P. (2019). *The Shroud of Turin. First Century after Christ!* Jenny Stanford Series on Christian Relics and Phenomena.

The Sudarium of Oviedo: Signs of Jesus Christ's Death
César Barta
Copyright © 2022 Jenny Stanford Publishing Pte. Ltd.
ISBN 978-981-4968-13-3 (Hardcover), 978-1-003-27752-1 (eBook)
www.jennystanford.com

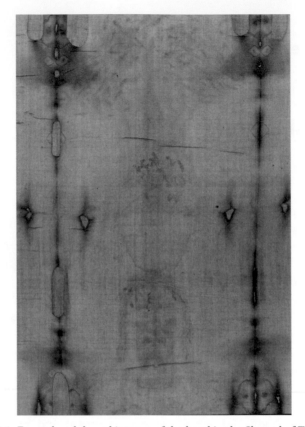

Figure 7.1 Frontal and dorsal images of the head in the Shroud of Turin.

However, for the purpose of this chapter, we will look at the part of the Shroud that covered the head, to allow comparison with the Sudarium that wrapped a head too. The Shroud has images of the frontal and dorsal sides of the head (Fig. 7.1). We are going to use this part of the Shroud to analyse the possible correlations.

7.1 Tradition

Something that is obvious when comparing the cloths is that both are preserved thanks to the tradition that considers them to be relics of Christ. But tradition alone is not enough to support their authenticity. For instance, there are other relics, considered by tradition to be Christ's, which certainly cannot be considered first-class relics belonging to or used by Jesus Christ.

In the Synoptic gospels the Shroud is mentioned although there is no mention of an image on it. Concerning the Sudarium, it is mentioned in the Gospel of St John as seen in the sepulchre, after the resurrection, with no other description but that it had covered the head of Christ and was left folded apart. And every object related to Jesus Christ mentioned in the gospels is likely to have been preserved ... or "recreated." However, these two relics are not easy to imitate. To reproduce the image of the Shroud has not been possible until today and trying to falsify the Sudarium of Oviedo would require in-depth knowledge of pathological and physiological processes along with the Jewish funeral rites involved.

Today, there is only one complete shroud that pretends to be the authentic one, located in the Turin Cathedral although there are many copies that are recognized as its reproduction. These copies only reinforce the authenticity of the original.

About the Sudarium of the Gospel, we saw in Chapter 2 that none of the other candidates, apart from the Sudarium of Oviedo, have any supporting evidence to be the authentic one.

The Shroud of Turin and the Sudarium of Oviedo, each of them, separately, have been analysed to determine whether they can be the sepulchral Shroud of Jesus of Nazareth and the Sudarium present in the tomb of Jesus, respectively. The assessment of scientific studies favours a confirmation of tradition rather than its rejection. Furthermore, if both showed features that correspond to the same individual, the authenticity of both would be reinforced.

7.2 Fabrics

One aspect of the comparative study is from the textile point of view. We have already described the Sudarium of Oviedo as a taffeta (or tabby) linen fabric. The Shroud is a twill or herringbone linen fabric. Table 7.1 presents some textile characteristics of both.

There are two remarkable aspects to highlight. First, the most significant similarity is the spun Z, which is not typical of Israel and may indicate the same dealer. Secondly, both fabrics show a relatively high density. However, the type of weaving of the Sudarium of Oviedo is coarser than the herringbone weave of the Shroud of Turin. The twill of the latter indicates a much more expensive fabric corresponding well with a Jewish burial purpose cloth.

Table 7.1 Textile characteristic of the Shroud and the Sudarium

Cloth	Shroud of Turin[a]	Sudarium of Oviedo[b]
Weave type	Herringbone twill	Taffeta
Fibre	Linen	Linen
Threads/cm in warp	38.3	43
Threads/cm in weft	25.8	19
Spun	Z	Z
Weight (mg/cm^2)	23	19.8

[a]Lemberg, F. (2005). *Le Tissu du Linceul de Turin. Traces d'une histoire très mouvementée.* MNTV n. 32. p. 33.
[b]Montero, F. (2004). *Descripción del Lienzo, química y microscópica.* Proceedings of the 1st International Congress of the Sudarium of Oviedo. Oviedo, 1994. p. 68.

As we saw in Chapter 6, Raymond Rogers analysed under his polarized light petrographic microscope threads of the Sudarium and threads of the Shroud. He concluded that there was a significant probability that the fabric of the Sudarium of Oviedo was related in time and place to that of the Shroud of Turin.

7.3 Type of Punishment and Death

It is worth comparing from the medico-legal point of view, the process that led to the death of the Man of the Shroud of Turin with the process that led to the death of the Man of the Sudarium of Oviedo.

7.3.1 Crucifixion

The features on the Shroud of Turin are abundant and unequivocally correspond to a Roman crucifixion of a male. In particular, the nail wounds on the feet and hands, the use of a *flagrum* (a particular Roman whip), the erosions of the shoulders, the wounds of thorns on the forehead and nape or the wound in the side, perfectly fit not only with a Roman crucifixion but also with the one that Jesus Christ suffered as described in the gospels. In that case it would be the crucifixion of a Jew, a peculiarity precisely concordant with a clean and prized shroud according to the Jewish custom at the time of Christ[2].

Evidence on the Sudarium for the type of punishment are not as accurate as those found on the Shroud. However, the forensic analysis indicated, with high reliability, that the person on whom the Sudarium of Oviedo was applied had been executed and remained in an upright position for more than an hour after death. We recall, from Chapter 6 on dating, that a vertical execution is only compatible with gallows, impalement, and crucifixion from an exhaustive list of death penalties[3]. The slow and soft effluvium of pulmonary oedema through the airways deduced from the Sudarium excludes both penalties of hanging and impalement. The gallows block the flow between lungs and mouth and nose. On the other hand, the pulmonary oedema is a consequence of heart failure that does not occur in impalement where internal bleeding predominates.

The Sudarium of Oviedo covered the face of a corpse of someone who died in analogous conditions to crucifixion and had also previously been mistreated to the point of having the hair soaked in blood. This is what this cloth shows. In short, the crucifixion is the execution that best fits the features presented by the Sudarium of Oviedo.

Moreover, the use of a sudarium to cover the face of the corpse perfectly fits in with the Jewish traditions. A crucifixion of an individual of any country under the Roman domination would have most likely ended with the corpse deposited in a mass grave without any kind of consideration. From the Roman point of view, it was not intended to use any sudarium to cover the face of a crucified person. On the contrary, in the crucifixion of a Jew, there was an urgent need to stop the bleeding since the blood had to be prevented from being lost. In fact, it was prescribed in the legislation of the Pentateuch and in the instructions of the Sanhedrin, to collect the blood and bury it

---

[2]Gitlitz, D. M. (2003). *Secreto y engaño. La religión de los judíos.* Junta de Castilla y León. p. 259.

[3]Barta, C. (2007). *Datación radiocarbónica del Sudario de Oviedo.* Proceedings of the 2nd International Congress for the Sudarium of Oviedo. pp. 137–158.

together with the executed. In this case, a Jewish bleeding corpse, with a bruised face, in which the fluid from the lung oedema could be seen leaking from the nose and the mouth, would undoubtedly have been recommended to use any type of cloth to cover the face of the executed and to retain the blood. This mandatory custom could have been performed in the time that Pilate took to grant permission to take the corpse for burial and pious Jews who tended to the crucified would have been in charge of this.

7.3.2 Scourging

Moreover, in the lower corner of the Sudarium there is a double stain which could possibly have been originated by a coup of the *flagrum*[4] (Fig. 7.2).

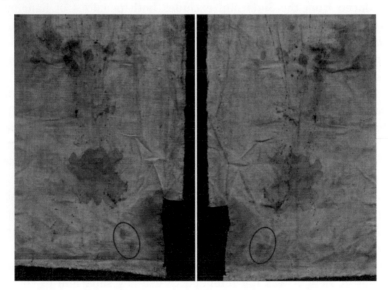

Figure 7.2 *Flagrum* double mark near the stain in the corner. It is clearer by the back side (left) than by the side in direct contact (right).

Attentive observation of the area under the "butterfly" shaped blood stain and next to the "corner" stain reveals two small dot-shaped blood stains that can easily be unnoticed. Furthermore, they do not seem to be related to the stains that surround them.

---

[4]Sánchez Hermosilla, A. (2012). *The Oviedo Sudarium and the Turin Shroud*. 1st International Congress on the Shroud. Valencia. pp. 55–58.

Look at the left image of the Fig. 7.2. The couple of small stains are separated from each other by 4.69 mm and their farthest points are at a distance of 21.42 mm. Their diameters correspond to 9.4 and 9.2 mm for the upper and the lower stains, respectively. If we look at the image on the right in Fig. 7.2, which corresponds to the opposite side of the cloth, the distance between them is 2.48 mm, and the distance between the extreme points is 21.15 mm.

The size of these stains is compatible with those that appear in the Shroud of Turin and with the *Flagrum Taxilatum* (Fig. 7.3) that is assumed to be the punishment tool used for scourging Jesus Christ when he was still alive.

Figure 7.3 *Flagrum* metallic edge compared to the double mark (left) and close up of the stains (right).

When the dot-shaped stains ("crown of thorns") of the Sudarium are superimposed over the correspondent group in the image at the nape of the Shroud, those stains that we assume to be *"flagrum"* marks in the Sudarium, fall under the hair in the Shroud and hence, cannot be seen in the Shroud on this part of the neck (Fig. 7.4). However, they are at the same horizontal level as the many perceptible *flagrum* marks on the shoulders of the Man of the Shroud. The *"flagrum"* marks on the Sudarium could therefore correspond to some wounds by the *flagrum* under the hair or also on the left shoulder of the Shroud. The reason is the almost sure displacement of this part of the hair and the corner of the Sudarium stitched to it when the Sudarium was tightened around the head because the pulling to the left. The corner of the Sudarium where the *"flagrum"* marks are could fall therefore some centimetres away from the position of the hair appearing in the Shroud. According to the reconstruction, the position of the hair wrapped in the corner of the Sudarium when it was tightened around the head was not the position seen in the Shroud but shifted to his left. After the removing of the Sudarium and in the absence of pulling, the hair could recover lightly the central position that it had before the use of the Sudarium as it appears in the image of the Shroud.

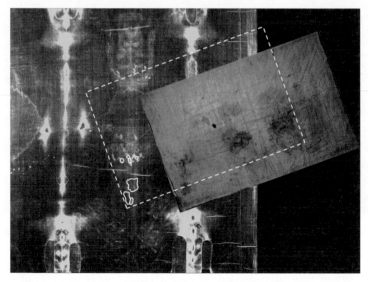

Figure 7.4 Superimposed contour of the Sudarium over the wounds at the nape of the Shroud of Turin and place of the *Flagrum* stains (red).

Concluding, this couple of small stains suggests the use of tool of a Roman penalty during the punishment undergone by the Man of the Sudarium, excluding the hypothetical crucifixion in the Islamic context and correlates with the *flagrum* marks of the Shroud.

7.3.3 Crown of Thorns

Traces of a crown of thorns punishment are also present in both cloths. The crown of thorns is an extremely specific torture of the Passion of Christ. At least we do not know of any other case apart from that of Jesus of Nazareth, in which prior to death on the cross, the individual suffered a similar torture. Both cloths show small clots originated by wounds produced by pointed objects clearly identifiable with those of a crown of thorns. These clots are located in the occipital area as well as in the forehead region. Specifically, they are evident in the area corresponding to the nape of both cloths (Fig. 7.5 right, for the Shroud and Fig. 7.6 bottom, for the Sudarium). We can also observe many of these clots on the forehead of the face of the Shroud (Fig. 7.5 left). However, in the forehead area, in the Sudarium, we can only identify one clot of this type (Fig. 7.6 middle).

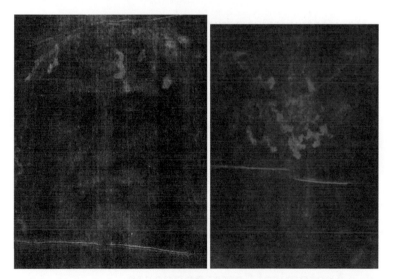

Figure 7.5 Clots of blood in the forehead (left) and nape (right) as traces of the crown of thorns in the Shroud of Turin.

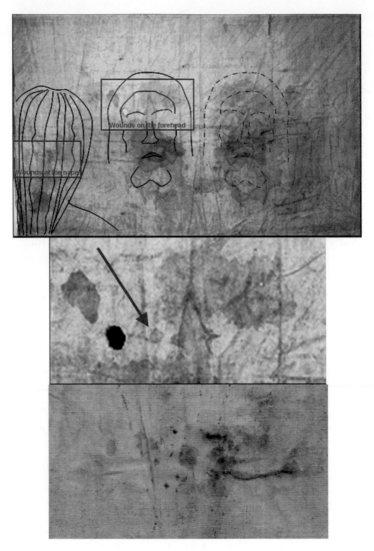

Figure 7.6 Top: Places of traces of the crown of thorns in the Sudarium of Oviedo for reference of the two lower images. Middle: Close-up of clot on the forehead. Bottom: Close-up of clots of blood at the nape.

7.4 Anatomical Correspondence

From the geometric point of view, we can compare the heads of the Man of the Shroud and of the Man of the Sudarium of Oviedo.

Pioneering work in this area was developed by Msgr Ricci whose first work concerning the Sudarium of Oviedo was published in his book *"L'Uomo della Síndone è Gesù"*[5]. He was the first one to deduce that the Sudarium had been folded with two layers in front of the face of the Man. He found an acceptable macroscopic fit between the face of the Shroud of Turin and the stains on the surface of the Sudarium. His conclusions were later improved by the work of the EDICES.

In Fig. 7.7, the face of the Man of the Shroud is superimposed in an increasing degree of appearance over the main stain of the Sudarium. It allows having a general sight of the anatomical coincidences.

Figure 7.7 Coarse progressive superposition of the face of the Shroud on the main stain of the Sudarium.

The main stain that was in direct contact with the face can be divided in three parts: the upper, the lower and the central

[5]Ricci, G. (1985), *L'Uomo della Sindone è Gesù*, Ed. Cammino, Milano. pp. 219–238.

one. Thus, we can see that the upper part corresponds to the forehead of the face of the Shroud, and the lower part to the mouth and finally, the narrower and central part of the stain that connects the upper and the lower parts, corresponds to the nose.

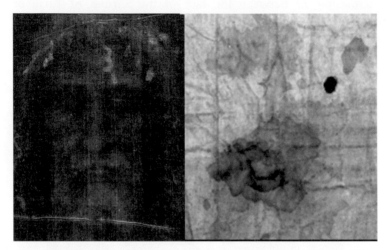

Figure 7.8 Face of the Shroud of Turin (negative) and the main stain of the Sudarium.

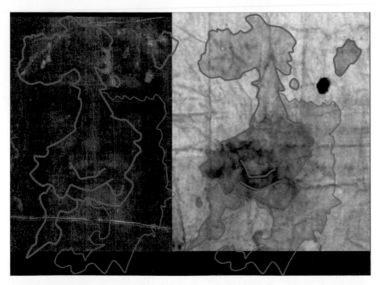

Figure 7.9 Outline of the main stain over the face of the Shroud of Turin and over the stain of the Sudarium.

In a more profound comparison, we look at the negative image of the face of the Shroud and at the main stain of the Sudarium that was in direct contact with the face (Fig. 7.8 and Fig. 7.9).

The comparison is made with photographs of both relics. It is important to highlight that we use the same scale for the image of both cloths. The first requirement is to ensure that the scale of these photographs is the same. Other comparisons have been published where this basic requirement was not met[6]. However, we must admit that some uncertainty remains in the scale when dealing with photographs. Thus, in order to reduce measurement errors, we started from photographs which showed the entire relics, that is, photographs in which their edges could be distinguished. Then, to obtain the scale, we measured the dimensions of each cloth on the photograph and compared with their best measurements of the real sizes. Once the scale is known, we can measure the distance between several prominent details around the area of interest on both fabrics. This approach has been further confirmed by taking photographs of a full-scale Sudarium facsimile which we located close to the face of a full-scale Shroud facsimile. In short, the divergences between the different approximations are always limited to the order of one millimetre.

The upper part of the main stain fits with a part of the forehead of the face in the Shroud. In the image of the Shroud, this area is limited by the edge of the hair, the eyebrows, and the clot in the shape of 3 (3 in the negative and the famous "epsilon" in the positive). These elements delimit a "reservoir" for the fluid that comes from the nose and mouth when the corpse was face down.

We highlight the widening or enlargement of the stain just at the place of the eyebrows where the capillarity of the eyebrows hair absorbed the fluid toward the interior of the eyebrows.

The lower part of the main stain in the Sudarium corresponds to the tip of the nose, the mouth, and the beard. When the images of both cloths are on the same scale and the forehead fits well, the contour of the stain which corresponds to the mouth and beard in the Sudarium should correspond to these anatomical

---

[6]Baima Bollone. *Risultatti della Valutazione dei Rilievi ... sul Sudario di Oviedo.* Proceedings of the 1st International Congress for the Sudarium of Oviedo. Fig. 8. p. 408.

elements on the image of the Shroud. However, it appears displaced downwards (Fig. 7.10). We will provide a possible explanation to this later in this chapter.

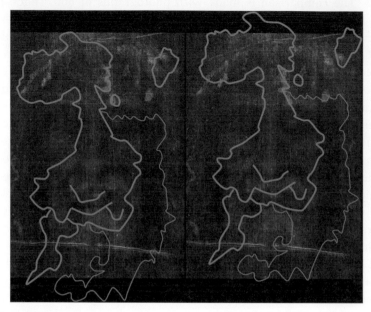

Figure 7.10 Upper and lower parts of the main stain do not fit simultaneously.

Now, we highlight that, if we correct this displacement, fitting the lower part of the stain of the Sudarium to the mouth of the image of the Shroud, we obtain a good correspondence on this area between both. First, the darker part of the stain in the Sudarium fits well with the nose and the mouth of the image which are the source of the fluid. In addition, both edges of the moustache are also observed in the Sudarium stain. Finally, the long clot that runs down from the left[7] corner of the mouth in the Sudarium also fits well with its corresponding clot on the Shroud. This clot was one of the main features that convinced Msgr Ricci that the Sudarium and the Shroud covered the same face.

In addition, there are other coincidences between the two cloths. Indeed, an obvious but undeniable common attribute is

---

[7]Left in the Fig. 7.10 that corresponds to the anatomical right side of the Man.

that in both cases the person is male because He has a beard and a moustache. In the case of the Shroud, the image offers all the evidence. In the case of the Sudarium of Oviedo, the presence of a beard and moustache is only deduced after reconstructing the stains by the forensic study. The distribution of the tracks fits not only with the presence of the moustache and beard, but also with a beard of the same bi-lobulate shape as that the Man of the Shroud presents.

7.5 Blood on the Face Area of the Shroud

Another difficulty when comparing the Sudarium and the Shroud is that in the Sudarium there is only blood, while in the Shroud, there is blood and also an image. Thereby, we should only compare the blood of both cloths. However, distinguishing between blood and other elements in the Shroud is not always easy. Indeed, an internationally broadcasted documentary assimilated a bloodstain in the forehead of the Man of the Shroud with a fabric hole in the Sudarium! While in the Sudarium of Oviedo it is relatively easy to distinguish blood stains, in the Shroud of Turin, in the area of the face, a previous treatment must be performed to separate blood from the image.

The procedure followed to highlight the blood on the face of the Shroud of Turin is not simple. The lifeblood that gushed from the wounds of the crown of thorns is easily identified. But it is no longer obvious that there are bloody stains in the beard area. The common visible-light photographs show that these stains around the mouth, if they exist, are much fainter and can be confused with the enigmatic image of the face. We have considered three approaches to better distinguish the possible blood on the face from the image in the Shroud:

- the image in the reverse side,
- the transparency image, and
- the numeric processing of the RGB (Red, Green and Blue) colour.

The reverse was the "hidden side" of the Shroud, that is, the surface covered with the reinforcing fabric. It has been analysed by various researchers[8] from photographs and scanner records. In Fanti et al.'s work, a mathematical correlation is used to avoid the subjective intuition and provide a more objective assessment. The conclusion in the paper is that a significant correlation between obverse and reverse sides of the fabric is only obtained in the facial area. Across the rest of the body image the correlation is negligible, with a slight exception in the area of the hands. Our interpretation is that the image does not penetrate to the reverse side and only appears on the obverse which was in direct contact with the body. But then, how can we interpret the correlation in the facial area? Well, it is verified that blood stains do penetrate from the obverse to the reverse side, as evidenced in the clots of the crown of thorns or in the wound on the side. Therefore, we think that the correlation can come from the fact that the face was much more loaded with bodily fluids and diluted blood substances than the rest of the body. These diluted fluids would have soaked the face and would have been held back notably by the presence of the Sudarium. Given the transparent nature of sweat and pulmonary oedema, their existence is not so evident on the obverse of the Shroud or surface exposed to contemplation. Therefore, the correlation between both sides in the face area in the Shroud comes from the fluid presence and not from the image. The fluids pass through but the image not. The use of the Sudarium in the area of the face makes the difference with the rest of the body. Across the rest of the body, evaporation was possible and thus sweat could be dried more efficiently as it was not covered by any cloth for hours until the body was taken to the sepulchre. The correlation found on the obverse and reverse in the face area would then have its origin in the body fluids that do not contain vital blood, but post-mortem

---

[8]Fanti, G., and Maggiolo, R. (2004). *The double superficiality of the frontal image of the Turin Shroud.* J. Opt. A: Pure Appl. Opt. 6. pp. 491–503. As R. Rogers observed, the colouring of the body image affects a maximum of three fibres in the depth of the threads and all the other fibres inside the thread do not have the typical colouration of the body image. Fanti assumes that the correlation in the face area is explained by few affected fibres on both side of threads. His Corona discharge hypothesis could explain image at both sides and nothing inside of the threads. Fanti, G. (2010). *Can a Corona Discharge Explain the Body Image of the Turin Shroud?.* Journal of Imaging Science and Technology 54(2).

blood retained by the Sudarium. As the Sudarium was removed just a few minutes before the use of the Shroud to cover the corpse, the fluids were still lightly wet and soaked the Shroud tissue. This would be a new indication of the use of the Sudarium of Oviedo by the Man of the Shroud.

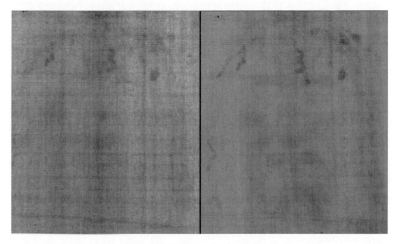

Figure 7.11 Reverse (left) and obverse (right) of the face area in the Shroud (© CIS).

If our theory is confirmed, the image on the reverse of the Shroud would allow us to distinguish the blood and hide the image (see Fig. 7.11)[9].

The second approach considered to separate the blood from the image is much more reliable and consists of the use of transparency photograph. The image of the Shroud shows a particular characteristic: in transparency photograph it is not perceived (the image disappears!) because it is not due to an added substance to the tissue that would block the light. If there is no blood, the light only must pass through the tissue. We see the image because of a fibre degradation, which reduces its reflectivity of the incident light and thus, it is only perceived by light reflection, but it does not affect the transparency. However, the blood stains are distinguished in the transparency photograph because they are substance added to the tissue that partially

---

[9]The image of the Shroud face in the Fig. 7.11 and following has been transposed right to left to facilitate the comparison.

blocks the path of the light. Here, we used the photograph taken by Barrie Schwortz in 1978 (Fig. 7.12), although compensating the heterogeneities due to the fringes of the twill. Unfortunately, the spotlight used for backlighting was located just behind the forehead and it created saturation in an important part of the image. To lighten this inconvenience, transparency photography has been combined with that of the reverse and the contrast has been increased in order to highlight the dark areas assimilated to blood. The result is shown in Fig. 7.13. It tries to show only blood stains while hiding the image. The dark areas would be due to the presence of fluids that impregnated the fabric. We remark that the areas of the nose, moustache, beard, and the left part from the cheekbone downwards show the greatest presence of fluids. These same areas correspond to the main central stain of the Sudarium of Oviedo that contained one part of blood to six of pulmonary fluid. Unfortunately, the area of the forehead is masked by the spotlight and cannot be compared as clearly.

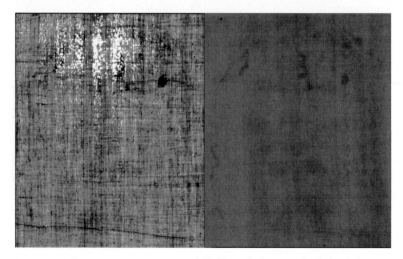

Figure 7.12 Transparency processed (left) and obverse (right) of the area face in the Shroud.

Finally, we use the numeric processing of the RGB colour provided by Dr Propp[10]. The image of the Shroud was split in

[10]Propp, K. of the Turin Shroud Center of Colorado Springs. U.S.A. The work was presented in the 2nd International Congress for the Sudarium of Oviedo in 2007 during the talk of Jackson J. *Preliminary Comparisons of Scientific data collected*

the three main colours: red, green, and blue calculating the score of the intensity of these colours in every pixel. Then, the value of the ratio red to blue was plotted against the value of the green. It is the plot at the left in the Fig. 7.14. In this plot, we selected the pixels corresponding to the known blood clots, such as the 3-shaped stain in the forehead and the ratios with values close to it are selected as blood pixels. It is the area marked inside the cloud of points in the graph. All the pixels inside this selected area have a similar colour composition as the true blood in the Shroud. Finally, we plotted in red these identified blood pixels in the image of the face indicating the possible presence of bloody fluid (Fig. 7.14 right).

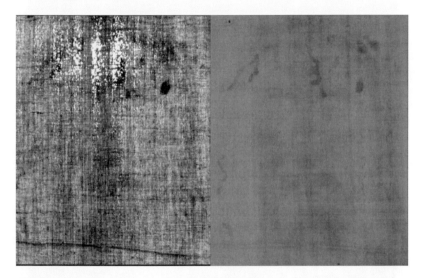

Figure 7.13 Transparency plus reverse (left) and obverse (right) of the area face in the Shroud.

To reinforce the trace of blood in the face of the Shroud, we again combine this last processed image with the previous image in which we added the reverse and transparency images (Fig. 7.15). The image improves the goal of showing blood presence and hiding the image.

from both the Shroud of Turin and the Sudarium of Oviedo. But it does not appear in the Proceedings. Figures granted by personal communication.

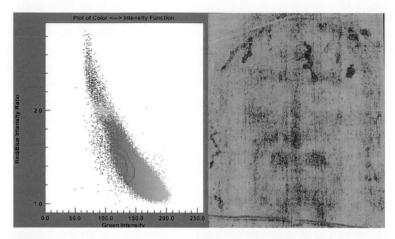

Figure 7.14 Plot of the ratio red/blue against green intensities (left). Green dots inside the inner outline correspond to blood. Dots inside the outer outline have a value of the ratio similar to the blood pixels and they should also have blood. At the right, the face in the Shroud with pixels in red that have a value of the ratio similar to the blood pixels.

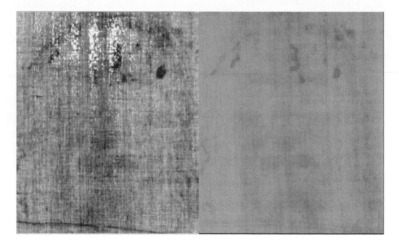

Figure 7.15 Combination of reverse, transparency and RGB processed images of the Shroud (left) and face of the Shroud (right).

By superimposing the outline of the central stain of the Sudarium on the face of the Shroud elaborated according to the criteria described (Fig. 7.16), a good correspondence can be seen of the forehead, the eyebrows, the 3-shaped clot, the drop near it and the upper part of the nose.

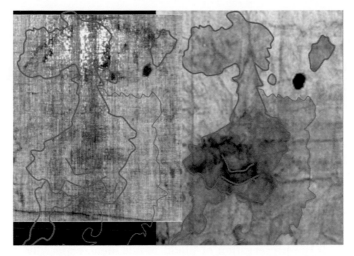

Figure 7.16 Outline of the main stain of the Sudarium (right) over the "blood" of the face of the Shroud (left). Matching forehead.

However, as disclosed above, the stain in the Sudarium that corresponds to the mouth and beard appears displaced downwards with respect to these anatomical elements of the face on the Shroud. To match these stains in both cloths, it is necessary to move upwards the outline of the main stain over the processed image of the Shroud (Fig. 7.17).

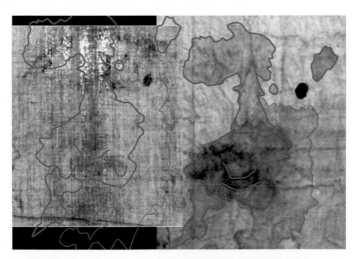

Figure 7.17 Outline of the main stain of the Sudarium (right) over the "blood" of the face of the Shroud matching the mouth (left).

A good correspondence can be seen now around the lower part of nose, mouth, moustache, and left cheekbone where both cloths show the maximum bloody fluid quantity. With the approach described, the clot coming down from the mouth at the lower left part of the outline is now better perceived in the Shroud. This clot is difficult, to distinguish in direct photographs. However, it convinced Msgr Ricci that the face which was covered by the Sudarium and by the Shroud was the same face.

With this procedure, we get clearer geometric correspondence of the bloody stains in both cloths. However, the blood correspondence is not only geometric.

Today we know the probable blood group on both cloths. Several blood group analyses have been performed in both cases[11]. Although the preservation of blood is not ideal, especially in the case of the Shroud, all the results have been unanimous— in the Turin cloth as well as in the Oviedo cloth —the blood belongs to group AB. This coincidence is already highly significant given the relative scarcity of this blood group, present in only 3.2% of the world population. This means that if we randomly selected 2000 individuals from humanity and grouped them together two by two, only in one of the 1000 couples would both individuals have group AB. We mean, there was roughly a one in a thousand chance that the blood on these two cloths matched. In addition, an interesting circumstance to be noted in this case is that among the Jewish population the percentage of AB group increases up to 18%, if we limit to the Jews of the north of Palestine.

On the other hand, while in the Sudarium, the area stained with blood is evident; in the Shroud a treatment has been necessary to highlight it. This shows that the Sudarium retained more blood than the Shroud of Turin. It should be pointed out that the arrangement of both cloths on the face that they covered was different and they were used in a different time. When the Sudarium was placed in front of the corpse's face, the corpse was in an upright position with the head tilted forward. The fluid

---

[11]Goldoni, C. (1994). *Estudio Hematológico sobre las muestras de sangre del Sudario tomadas en 1978*. Villalain, J, D. (1994). *Naturaleza y formación de las manchas*. 1st International Congress on the Sudarium of Oviedo. Universidad de Oviedo. Oviedo. Baima Bollone P. L. et al. (1982). *Identificazione del gruppo delle tracce di sangue umano sulla Sindone*, Sindon, Quaderno No. 31, Dicembre 1982, pp. 5–9.

that slowly emanated from the nostrils and mouth tended to fall by gravity and soaked the Sudarium that had been placed there precisely to collect the fluid. However, the Shroud was stained hours later only with the fluid remaining on the face oriented upwards. In the case of the use of the Sudarium, the head was tilted down, and then, most of the time, the fluids that exited through the facial orifices slid towards the tip of the nose and through the corner of the lips, affecting the most prominent parts of a limited face area. This would account for the greater amount of blood observed in the Sudarium with respect to the Shroud.

Now we deal with the impossibility to fit the outline of the main stain of the Sudarium simultaneously on the mouth and on the forehead of the face of the Shroud. As we saw, if the stain match well on the forehead, we must move upwards the outline of the main stain over the processed image of the Shroud to match the area of the mouth.

Faced with this mismatch, one would tend to think that there is a problem of scale. However, the scale has been checked and verified and the discrepancy largely exceeds the estimated margin of error. Well, we believe that the explanation is in the different adjustment that the Sudarium and Shroud experienced on the face they covered. The forensic reconstruction of the use of the Sudarium shows that it was closely fitted to the profile of the face by sewing it to the strands of hair and beard and, at times, compressing the area of the nose and mouth with a hand. However, the Shroud was settled gently on the face keeping contact only with the most prominent parts of it, such as the tip of the nose and chin. These differences imply that the projection of the base of the nose that goes from the tip of the nose to the upper part of the upper lip did not occupy space on the Turin cloth since it is projected as just a dimensionless line. On the contrary, a part of the Sudarium was tightened against the entire length of the base of the nose when an edge was sewn to the hair of the nape and the opposite fold was sewn to the hair and beard of the right side. When unfolding the Oviedo cloth, the mark on the tip of the nose is separated from the mark on the lip by the distance corresponding to the base of the nose (Fig. 7.18). The typical Jewish nose is prominent enough!

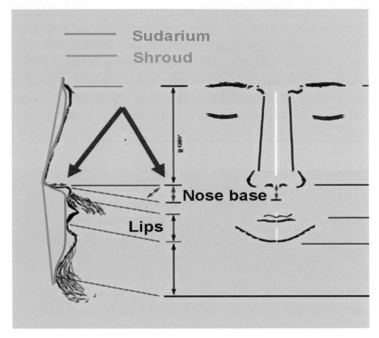

Figure 7.18 Difference of placement of the Shroud and the Sudarium on the nose area.

The length of the area of the Sudarium that was covering the face would be greater than the equivalent area of the Shroud due to the better fit of the Sudarium to the face. This should be the reason for which the mouth and forehead of both cloths cannot be simultaneously superimposed. The shift is produced by the base of the nose which is partially present in the Sudarium and absent in the Shroud. If we removed the gap of the stain corresponding to the base of the nose, we can match simultaneously forehead and mouth (Fig. 7.19). This gap is around 1.2 cm which is coherent with the estimation of Dr Miñarro[12] who proposed 3 cm for the projection of the nose as a result of his simultaneous studies of the Shroud and Sudarium.

We can now analyse a drop on the forehead near the eyebrow at the right in the previous figures[13] one of them reprinted in the Fig. 7.20 with an arrow indicating this drop.

---

[12]Miñarro, Juan M. (2012), *Plastic Reconstruction of the Man of the Shroud*. 1st International Congress on the Shroud. Valencia. p. 26.
[13]The left eyebrow of the Man.

Figure 7.19 Once the gap corresponding to the base of the nose (left ©CES) is removed, the stain matches simultaneously with forehead and mouth of the face of the Shroud (right).

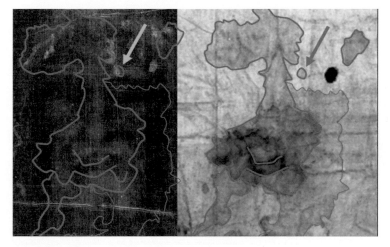

Figure 7.20 The arrow indicates the isolated drop in the forehead of the Sudarium.

This is a stain in the forehead identified in the Sudarium and associated with the crown of thorns. It is isolated with respect to the rest of the main stain. The first fact that stands out is that the entire central stain on the Sudarium forms a continuum due to the bloody oedema coming out of the nose and mouth and spreading millimetre by millimetre until reaching the upper edge of the forehead when the corpse was face down. However, the drop on the left eyebrow on the Man seen in Fig. 7.20 is close to the entire main stain but separated from it without continuity. The drop has not reached there dripping from the nose or mouth like the rest of the stain. It could, however, come from one of the puncture wounds in the forehead. This drop has its correspondence with a drop in the Shroud. To see it, we superimpose the nose; the frown and the forehead in the overlap of the Sudarium and Shroud. Then, the lower drop of the 3-shaped clot seen in the Shroud falls exactly in the same position on the eyebrow in both cloths. In addition, when this drop is amplified on the Turin cloth, an irregularity is observed with respect to the rest of the "three-shaped" clot. The central part of the drop is absent in the Shroud!

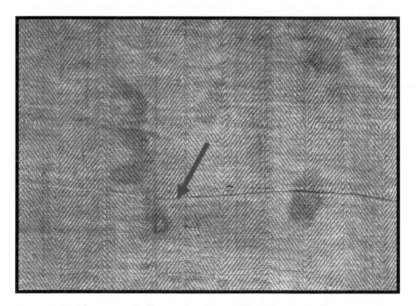

Figure 7.21 The arrow indicates the drop without its central part.

The rest of the clot and the other small spots are complete. To appreciate this detail, Fig. 7.21 uses the image of the scanner made in 2002 during the restoration and provided by the

International Centre for Sindonology in Turin (CIS). We may suggest that the missing part of the Shroud's drop remained in the Sudarium when it was removed.

There are other isolated stains much larger than the one just mentioned that appears in the upper right area of the forehead in the photographs of Fig. 7.22. Their position falls close to an accumulation of blood on the side of the forehead of the Shroud that becomes more obvious when using the processed photograph to highlight the blood. They are outlined in blue in Fig. 7.22.

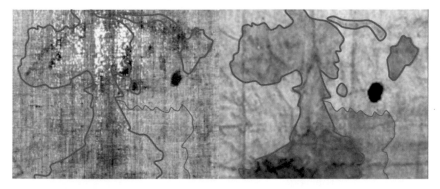

Figure 7.22 Displaced stains outlined in blue.

These stains can be found in the Sudarium and in the Shroud and they have a similar shape. However, they are displaced when the main stain and the isolated drop are well superimposed on the forehead. If both cloths had wrapped the head in the same way, we expected that the stains would overlap. However, this is not the case. A possible explanation is that the head is not flat, and these stains are in areas where the head starts to curve. Again, we suggest that the Shroud covered the head on an almost flat shape, while the Sudarium adjusted to the development of the head that it enveloped. This discrepancy raises an additional question: Were these stains projected on the Shroud or did they soak the fabric by contact? The clots on the side strands of hair in the Shroud look like they were soaked by contact and appear displaced from the image of the strands. In any case, we should not expect a perfect fit between every stain from the Shroud and the Sudarium because of the different way they were used and of the slight but certain displacement of the Sudarium along the different arrangements it undergone during its use. We must not forget that the Sudarium was knotted at the upper part of the head and wrinkles between stains increase the original distance between them when the cloth is deployed.

7.6   Correspondences in the Back of the Head

Now, we proceed to compare the occipital area and the rear lock of hair. In this area, in both cloths, there is a group of clots correspondent to small wounds clearly identifiable with those produced by a crown of thorns. Although the crown covered upper and lower part of the scalp, most of the stains appear at the lower line. This area of the head is where the head reverses its slope. Then, if a drop slipped by the scalp in the upper part of the head, it would descend by gravity adhered to the scalp. But when it reached that area of the neck, where the curvature of the head is tangent to the vertical, the drop tends to fall in vertical, crossing the layer of hair and going outwards. It could be the reason why more clots appear in that area than in the upper area of the head.

A cursory glance at the occipital area of both cloths, Sudarium and Shroud, already provides the impression that this entire area of pinpointed stains in the Sudarium of Oviedo is fully inscribed in the occipital area shown by the Shroud of Turin. In addition, it has been proved that in both cases the stains are life blood.

We proceed then to fit the stains after taking care of the scale. In the first attempt, a transparency of the Sudarium was superimposed on an image of the Shroud maintaining the original portrait orientation of the photographs. This way, it was not possible to find a convincing overlap. After overcoming the disappointment and in a second attempt, we discovered that rotating the transparency by 19º approximately, the fit was very satisfactory (Fig. 7.23). This rotation would be a consequence of the Sudarium being wrapped around the head. The natural tendency to maintain the orientation of the photos is understandable, but nothing prevents the Sudarium from turning when wrapping the head and this would be even more logical.

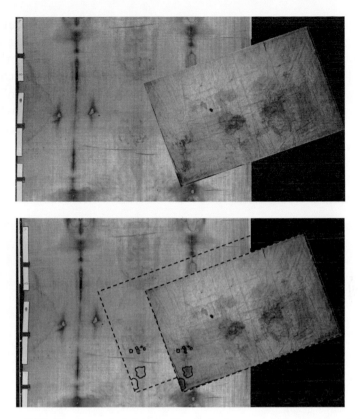

Figure 7.23 Overlap of stains in the Sudarium over the stains of the Shroud in the occipital area.

The correspondence has been verified with computer image processing. The maximum correlation between the photographs of these areas of the two cloths has been sought, allowing the variation of four different parameters: displacements in x and y coordinates, rotation, and the scale factor[14]. The computer has objectively found two maximums of common information but only one of both maximums respects the same scale between the two images. In addition, in the common scale solution, the angle of rotation was confirmed to be around 19º.

In an attempt to graphically show the coincidence, we combined photographs of both cloths one in red and the other in green (Fig. 7.24). So that when superimposing both coloured images (Fig. 7.25), parts that do not coincide show their original colour and those that coincide are darker. That is, the overlay is distinguished in darker colour highlighting evident matches between both. For example, both have a main stain of a very similar size and shape that is flanked by other smaller stains at a similar distance. We can count in that area of the Shroud about eight clots of which six have their corresponding in the Sudarium. That is to say, in the crown of thorns, which is an exclusive characteristic of the Passion of Christ, we obtain a 75% agreement between both relics.

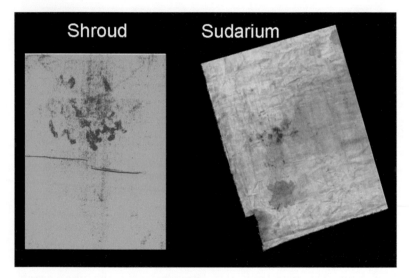

Figure 7.24 Green image of the nape of the Shroud (left) and red image of the corresponding area of the Sudarium (right).

Possible discrepancies in this occipital area may come from the different arrangement of both cloths on the nape of the subject. In the case of the Shroud, the back of the head is left on the cloth spread over the slab without any significant wrinkles. In the case of the Sudarium, the cloth was wrinkled in that area, because it was sewn to the hair as indicated by the row of stitches and the butterfly-shaped stain. In addition, it was folded along the middle of this butterfly stain (see Chapter 3).

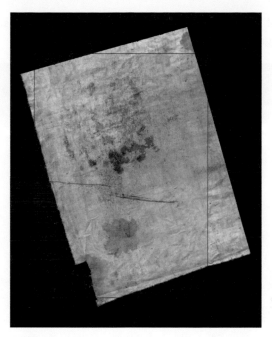

Figure 7.25 Overlap of green and red images of Shroud and Sudarium to highlight coincidence of stains.

On the other hand, in both cloths, these stains which come from wounds caused by pointed objects show characteristics of vital blood, that is, that they flowed from the subject while he was still alive. It should be noted the uniqueness of this type of vital blood in the Sudarium where most of the bloody material detected emanated while the Man was already dead. And precisely in this area, there is the same type of vital blood in the Shroud like in the Sudarium (Fig. 7.26).

In summary, the stains in the occipital area of the Turin cloth and of the Oviedo's cloth show a remarkable correspondence in size, relative position, and genesis as both come from vital blood.

Another coincidence between both cloths concerns the tuft of hair in the back of the neck. Both subjects have long, matted hair pressed back like a ponytail at the back of their head. This fact is evident in the negative image of the Shroud of Turin with reinforced contrast. A long strand that falls from the nape to the gap between shoulder blades is clearly visible on the dorsal part (Fig. 7.27). The tuft resembles a loose ponytail with no signs of fastening.

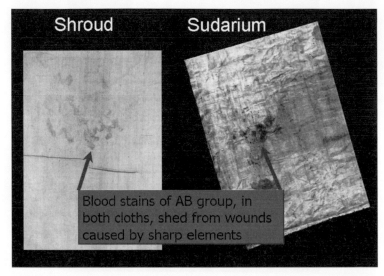

Figure 7.26 Stains of live blood of AB group in both cloths in the nape area.

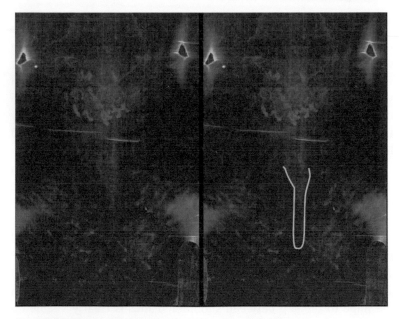

Figure 7.27 "Ponytail" at the back of the Man of the Shroud.

This peculiarity has often been attributed to a typically Jewish hairstyle from the time of Jesus Christ. We believe that there is a simpler and more credible explanation—that it is also a consequence of the use of the Sudarium of Oviedo on the head of the Man of the Shroud, as we will see later.

On the other hand, in the Sudarium, the butterfly shaped stain located in the lower left corner, appears just below the dot shape stains of the crown of thorns on the nape and has been reproduced in the laboratory by pressing a cloth against a bundle of cotton stained with blood (Fig. 7.28). The bundle of cotton simulates specifically the lock of hair. Therefore, in the Sudarium, this stain could also correspond to a "ponytail".

Figure 7.28 "Butterfly" stain and the laboratory simulation.

7.7   Coincidence for Dust on the Tip of the Nose

The subject of this section is based on the X-ray fluorescence finding[15] described in Chapter 3. We identified particular spots

[15]Barta, C., et al. (2014). *New Coincidence between Shroud of Turin and Sudarium of Oviedo*. Workshop on Advances in the Turin Shroud Investigation (ATSI). September 4, 2014.

in the Sudarium with the highest quantity of dust located near tip of the nose. Those spots show a statistically significant higher presence of dust which would have been fixed by the physiological fluids while they were still fresh. The X-ray detector revealed the presence of calcium, a mineral associated to soil dirt. On the other hand, an unexpected excess of dirt around the tip of the nose was also detected in the Shroud of Turin. This could be other astonishing coincidence between both cloths. The left of Fig. 7.29 shows the location of the three dustiest spots close to the tip of the nose referred to the main stain of the Sudarium and superimposed on the face image of the Shroud. The right of Fig. 7.29 shows the plot of the calcium quantity recorded from the X-ray fluorescence carried out on the Shroud[16]. In Fig. 7.29, both results are drawn on the same Shroud image for comparison purpose. The excess of calcium is close to the tip of the nose in both cases.

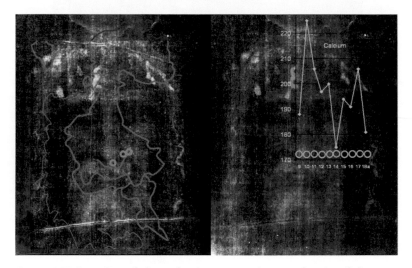

Figure 7.29 Location of three dustiest spots close to the tip of the nose in the Sudarium (left) and intensity of Calcium on the face of the Shroud (right). Elaborated from Morris et al.

Literature concerning the investigation of the dirt of the nose in the Shroud of Turin is reproduced hereafter:

---

[16]Morris, R. A., et al. (1980). *X-ray fluorescence investigation of the Shroud of Turin.* X-Ray Spectrum. vol. 9, no. 2. p. 44. Table I.

"Detailed photographs and microscopic studies of the cloth in the nose image area show scratches and dirt. These are consistent with the nose having made contact with the ground, most likely as the result of a fall"[17].

"Visual observation of the heel area at 500 times magnification revealed the presence of very fine yellowish particles suggesting dirt; the nose area might also contain dirt or residual skin material"[18].

"Pellicori and Evans noted significantly higher concentrations of particulates in the nose and foot regions of the image. In these areas, X-ray fluorescence indicated statistically significant excesses of iron above background levels"[19].

The particular accumulation of soil dirt in this area of the face was attributed to the result of falls in which the nose touched the ground.

The coincidence in the high concentration of dust next to the tip of the nose in the case of Sudarium and Shroud could be considered exceptional. However, we have to answer the question whether it is exceptional or rare to accumulate more quantity of dust at the tip of the nose as compared to the rest of the face and body, or, on the contrary, is it a common case. Only in the former case, this coincidence would indicate that something unusual has happened for finding it in Sudarium and Shroud and it would be a decisive sign of identity between the face of both cloths. In contrast, in the latter case, if it is common that, in a dusty environment, more dust is accumulated on the nose as compared to the rest of the face, this coincidence would not be conclusive.

To assess this issue, some tests were carried out staining a face with dust and evaluating its distribution in different parts of the face. The setup consisted in placing a sieve with different types of dust about 45 cm above the vertical of a fan that propelled the dust upwards at an angle of about 45° with respect

[17]Bucklin, R. (1982). *The Shroud of Turin: Viewpoint of a Forensic Pathologist*, Shroud Spectrum International. no. 5 Dec 1982. pp. 3–10.

[18]Pellicori, S., and Evans, M. (1981). *The Shroud of Turin through the microscope*, Archaeology, vol 34. Jan/Feb, pp. 35–43.

[19]Schwalbe, L. A., and Rogers, R. N. (1982). *Physics and Chemistry of the Shroud of Turin: A Summary of the 1978 Investigation*. Anal. Chim. Acta. vol. 135, no. 1, pp. 3–49.

to the horizontal to fall in a parabolic trajectory (Fig. 7.30). Dust was simulated in successive tests by wheat flour, plaster and an orange colouring product composed of corn flour, tartrazine, and salt. The face was placed trying to receive the dust stream in a path close to the horizontal where the dust spread and reached its maximum height.

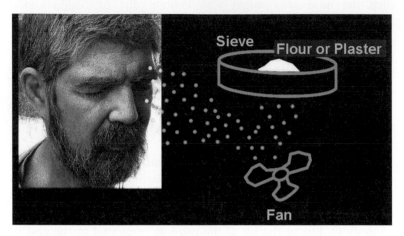

Figure 7.30 Setup to spread dust on a face.

Successive photographs were taken at intervals of about 30 seconds as the concentration of dust increased on the face. Once the stream of dust (flour, plaster, and coloured product) finished, an exhaustive photograph record was made of all the areas of the face. Then, we used image processing to analyse the photographs. Flour and plaster allowed a reliable analysis of the particles on the skin but were difficult to distinguish from grey beard and hair. However, the orange colouring product made possible to better distinguish the particles on the hair.

We selected 4 photographs of the flour, 5 of colouring and 8 of plaster for the image processing. Then, three rectangles of 264 × 204 pixels were selected in each photograph representing areas from the nose, the cheekbone, and the forehead (Fig. 7.31). Each rectangle was then divided into four quarters of 132 × 102 pixels. This size approximately corresponds to the measurement window of the X-ray fluorescence detector used in the Sudarium of Oviedo.

Figure 7.31 Three areas selected for the image processing.

Thus, we analysed 4 quarters from 3 rectangles of 17 photographs which summed up to 204 image processing. The procedure was as automatic as possible to avoid subjectivities and it was applied in a systematic way. The image processing was performed by the particle counter script of ImageJ 1.43u[20].

The maximal concentration of "dust" on the nose was found in less than half of the 17 cases analysed. However, the average of the cases showed preferential accumulation of dust quarter 3 of the rectangle of the nose (Fig. 7.32). It corresponded to the lobe of the nose near the spot where the maximal concentration was found in the Sudarium and the Shroud. The following dustiest spots in the experiment corresponded to the cheekbone, which has a concentration quite similar to that of the lobe of the nose. Although the accumulation of dust near at the tip of the nose shows slight preference, we remark that differences among different parts of the face were not statistically significant in any case, contrarily to the case of the sacred cloths.

---

[20]Particle counting seminar Severo Ochoa Molecular Biology Center and the Autonomous University of Madrid. http://babia.cbm.uam.es/~smoc/Software_Manuales_SMOC/Seminario_ImageJ-contaje_de_particulas.pdf.

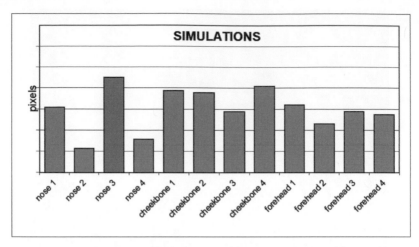

Figure 7.32 Concentration average of dust in the selected frames of the face resulting of the experiment.

Thus, our results show that the specific area of the face, i.e., the lobe of the nose that received the maximal concentration of dust in the Sudarium and Shroud often could receive the maximal quantity of dust in a common dusty environment. However, the concentration in this spot does not stand out over the rest; the differences were small and not statistically significant. Contrarily, the difference is significant in the case of the sacred cloths. In the case of the Sudarium, the X-ray fluorescence of the spot close to the tip of the nose stands out according to the statistical criteria. It is technically an outlier even if we compare this value with the following 11 points that present the highest concentration of calcium in the entire canvas (Fig. 7.33).

The same particularity occurs in the case of the Shroud. If we apply the statistical criteria of outlier[21] to the seven X-ray fluorescence values obtained in the face, the spot close to the tip of the nose statistically stands out of the group (Fig. 7.34).

Thus, in both cloths, the accumulation of dust at the point of the nose is certainly exceptional indicating that something particular happened to justify this accumulation. Possible explanation could be a fall face down against soil. In any case, this coincidence supports that both cloths were used on the same face.

---

[21]Chauvenet criterion.

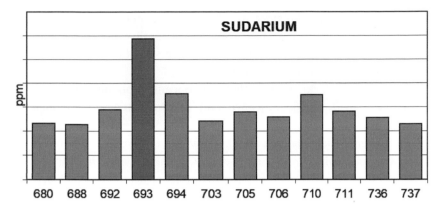

Figure 7.33 Concentration of dust on the 11 spots with the highest concentration of calcium in the Sudarium.

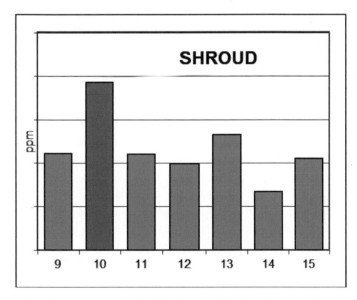

Figure 7.34 Concentration of dust in the 7 spots of the face in the Shroud.

If the reconstruction of the events proposed by EDICES is correct, the Sudarium would be closely fixed on the face and the most prominent parts would come into intimate contact with the skin and its contaminants. Particularly, in the area of the mouth and nose, a strong pressure is deduced, probably by a hand, which was intended to contain the nosebleed. This would bring the part of the nose and the inner cheekbone into close contact with the fabric. The area of the cloth around the nose would have had a strong transfer of face dust and that area would have transferred dirt in a similar way.

Finally, we acknowledge that the concentration measurements obtained by X-ray fluorescence and the concentration obtained by imaging are of a different nature. In addition, the data from our experiments tell us how the powder is distributed directly on the face, while in the case of the Sudarium and the Shroud; it involves the transfer of the powder from a face to the fabric that covers it. For such a transfer, even minimal contact between skin and cloth would be necessary. The pressure of the contact would also have influence. The higher the pressure is, the higher the transference is. However, our approach allows comparing the statistics of the data.

7.8 Reconstruction of the Head Provided by the Shroud and the Sudarium

Dr Juan Manuel Miñarro[22], performed the three-dimensional reconstruction of the body of the Man of the Shroud. In particular, he worked on the reconstruction of the head. His results were first presented as a communication for the second congress of the Sudarium in 2007[23] and later as a full contribution to the 1st International Congress on the Shroud held in Valencia[24]. For his study, he took data from both cloths: The Shroud and the Sudarium.

Before becoming aware of the existence of the Sudarium, he had started with images from the Shroud, and he created first and second versions of the three-dimensional head of the Man

---

[22]Doctor of Fine Arts. Director of the Department of Sculpture and History of Plastic Arts at the Faculty of Fine Arts of Seville. Member of EDICES.

[23]Miñarro, J. M. (2007). *Three-dimensional anatomical reconstruction based on the Sudarium of Oviedo and the Shroud of Turin*. Communication. Proceedings 2nd International Congress for the Sudarium of Oviedo. pp. 691–714.

[24]Miñarro, J. M. (2012). *Plastic Reconstruction of the Man of the Shroud*. 1st International Congress on the Shroud. Valencia 2012. A similar reconstruction has been carried out by Sergio Rodells. Real Colegiata de San Isidro. Madrid. Exposición esculturas de Sergio Rodella (Italia). OSMTJ-OSMTH. October 2021.

of the Shroud (Fig. 7.35). With the complementary study of the Sudarium, he created the third version of the tortured face.

Figure 7.35 3D model of face obtained from the Shroud stained with pigment and oil according to the Sudarium.

He performed different experiments to correct the effects of transformation from two dimensions to the third dimension and vice versa. The methodology was to place a fabric and an acetate on the previous 3D plaster model of the face of the Shroud. Then he marked in each case, on the outer surface, the positions of the main points of anatomical references.

After extending the fabric and the acetate on a plane, the references were perfectly consistent with the marks of the Sudarium of Oviedo. Major distortions occurred in the segments located below the tip of the nose, due to nasal step. The gap was estimated to be of 3 cm in the experiment. And between the tip of the nose and the chin, the distortion was found to be 2 cm, which could imply slight opening of the mouth or small lip dropping.[25]

---

[25]Rodella, S, arrives at the same conclusion. Rodells, S. Real Colegiata de San Isidro. Madrid. Exposición esculturas de Sergio Rodella (Italia). OSMTJ-OSMTH. October 2021.

For further research, the 3D face model was stained with a mixture of pigments and linseed oil, and then it was covered with a soft fabric, fitting it especially on the oral-nasal area, nasal bone, cheekbones, brow, and forehead. Thus, the fabric was stained with the paint mixture (Fig. 7.36).

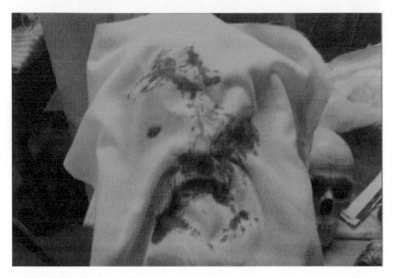

Figure 7.36 A cloth over the 3D model received the stains of the model.

After removing and unfolding the cloth, the morphology of the stains in two-dimensional extension was extremely similar in shape and arrangement to the original main stain of the Sudarium.

The conclusion of these experiences is that blood stains present on the three-dimensional face based on the Shroud are consistent with the two-dimensional marks of the Sudarium as there was an almost full coincidence between the proportions of the human face of the Shroud and the Sudarium.

The experiment allowed locating anatomical landmarks on the Sudarium shown and labelled in Fig. 7.37. The following numbers indicated the anatomical features:

1. *Glabella.* Between the eyebrows and above the nose
2. *Nasion.* Depressed area between the eyes
3. *Rhinion.* Join of the internasal suture and cartilage

4. Nasal spine. Base of the nose
5. Chin

Through this experiment, it was verified that the central stain of the Sudarium probably comes from a real face with the specific anatomical characteristics of the Man of the Shroud.

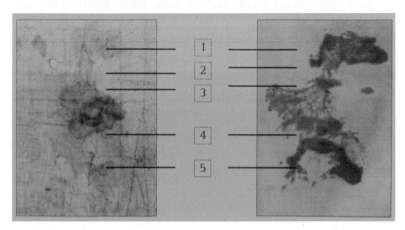

Figure 7.37 Correspondences of the anatomical features with the areas of the main stain of the Sudarium. Numbers explained in the text.

7.9   Summary of Coincidences

The Sudarium of Oviedo and the Shroud of Turin are two cloths with very different trajectories. However, tradition has regarded both as relics of Christ. Each of them, separately, has been analysed to verify the possibility of being sepulchral cloth. The result of scientific studies favours a confirmation of tradition.

Both fabrics are Z-spun, and they are related in time and place according to the Raymond Rogers analysis.

Both cloths were used on a male individual who had died in an upright position full compatible with crucifixion, after having suffered torture that produced severe pulmonary oedema. While there is evidence of scourging in the Shroud, the Sudarium shows a mark that is also compatible with scourging. Both individuals have endured a recognizable torture such as the crown of thorns.

Both cloths have been used on an individual with common anatomical features. The Man had beard, moustache, and long hair. A good correspondence is also observed between various anatomical elements of the face, such as the nose with its nostrils and fins, the brow ridges, the size of the mouth and chin, and even the shape of the beard. He had also in both cases long hair and at his back resembling a ponytail.

Both individuals shed bloody fluid from their noses and mouths. Once the presence of blood in the face of the Man of the Shroud is highlighted, a good correspondence can be seen among many stains, in particular, the clot that convinced Msgr Ricci of the identity of both faces coming down from the mouth. The edge of a drop near the "epsilon" on the forehead remained in the Shroud while the central part remained bonded to the Sudarium.

The coincidence of the blood group, which in both cases is group AB, is very significant since the probability of coincidence is approximately one in a thousand.

The distribution of stains of vital blood in the nape area corresponding to the crown of thorns coincides in 75 per cent of its geometric arrangement.

In addition, both cloths show an accumulation of dust in the part that covered the nose.

With a 3D model of the head of the Man of the Shroud it was verified that the central stain of the Sudarium should have come from a real face with the same specific anatomical features of the Man of the Shroud.

It is extremely unlikely that all of these coincidences occur by chance. Tradition could very well be right.

Chapter 8

Was the Sudarium Used on the Man of the Shroud?

In this chapter we will deal with two issues in the Shroud that require some explanation because of their inconsistency: The claimed "ponytail" and the Man's hair remaining next to the cheeks. They can have their origin due to the use of the Sudarium on the individual who was wrapped later in the Shroud. That is, we wonder if the Sudarium of Oviedo was used on the Man of the Shroud.

8.1 Ponytail

We first deal with the presence of the tress hair looking like a ponytail worn by the Man of the Shroud[1]. In the previous chapter we mentioned evidence of this feature in both cloths. The first coincidence is related to the form of the hair. On the back of the Man of the Shroud the hair is apparently arranged in a "ponytail" shape. It falls between the shoulder blades down to

---

[1]Barta, C. (2007). *The Sudarium of Oviedo and the Ponytail on the Shroud.* Shroud Newsletter, n. 66, December 2007. pp. 2–9. Barta, C. et al. (2014). *New Coincidence between Shroud of Turin and Sudarium of Oviedo.* Workshop on Advances in the Turin Shroud Investigation. September 4, 2014.

The Sudarium of Oviedo: Signs of Jesus Christ's Death
César Barta
Copyright © 2022 Jenny Stanford Publishing Pte. Ltd.
ISBN 978-981-4968-13-3 (Hardcover), 978-1-003-27752-1 (eBook)
www.jennystanford.com

the half of his back. This hairstyle has often been attributed to a typical Jewish fashion at the time of Christ[2]. However, it seems unlikely that the hair remains well arranged after undergoing the torture observed directly in the Shroud. Moreover, there is no evidence in the image of any artifact holding the tuft in such a shape. How could it have been held thus? A simpler and more probable explanation is provided:

> The so-called "ponytail" is the result of the use of the Sudarium of Oviedo which was placed and sewed around the hair in this area.

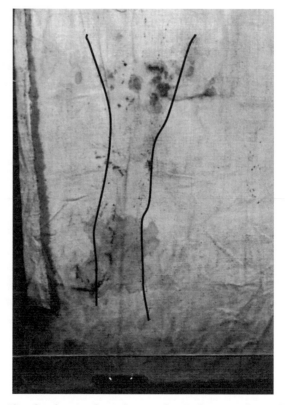

Figure 8.1 Sketch of row of stitches in the Sudarium of Oviedo running from the dot shaped bloodstains to the butterfly stain that delimit a lock of hair.

---

[2]Wilson, I., and Schwortz, B. (2000). *The Turin Shroud: The Illustrated Evidence.* London: Michael O'Mara Books. p. 42. Stevenson, K., and Habermas, G. (1998). *Dictamen sobre la Sábana Santa de Cristo*, 3ª ed. Planeta. p. 49.

Such is what we will attempt to show in this section. As we had explained in previous chapters, the edge of the Sudarium was placed at the level of the cervical vertebrae covering the back of the neck and the top of the head of the victim. The rest of the Sudarium covered the left ear, the nose and mouth and reached the right cheek. It was then folded back again on the nose and mouth. The right ear was left uncovered in the first phase. To hold the cloth in this position it was sewn with linen thread to the hair on the back of the head. Some threads remain in the Sudarium today. The seam can be deduced by two relatively parallel lines of holes where the needle went in and out. The seam ran from the dot shaped bloodstains to the butterfly stain (Fig. 8.1). The sewing holes cover not only the area between the dot shaped stains but also the stains themselves and then they go a few centimetres further on each side.

The forensic analysis concludes that the hair bundle remained between the fabric and the stitches tied up in this way for about two hours. Dirty hair soaked with blood and sweat and retained in this position for two hours preserves the given shape afterwards. This was experimentally verified using volunteers with long hair.

For a first approach a cloth of the same size as the Sudarium of Oviedo was sewn to the long hair of a girl (the author's daughter) in the same area of the nape as it was determined for the Spanish cloth (Fig. 8.2). The hair had not been washed recently but was not too dirty. Some hairspray was applied to simulate the caking effect of the dried blood. Once the cloth was unstitched and removed, we verified that the hair remained caked, thick, and compact near the side where it was squeezed by the seam at the edge of the fabric (Fig. 8.3). The shape of the tress caused by the fabric was maintained after its removal.

For more accurate simulation, new experiments were performed to simulate the process which the Jesus' hair underwent during His Passion. In one of the new experiments, the volunteer was the other daughter of the author and on another experiment the volunteer was a male with long hair and a beard. At the beginning of the experiment, the volunteers had clean hair and no perfume products. Thus, the blooding sweat experienced in the Mount of Olives was simulated with 25 cc of physiological serum mixed with 12.5 cc of blood drawn from the veins a few moments before by a nurse. To apply this mixture to many points on the scalp, a dispenser bottle with a dropper was used. Hot air was then applied for 5 minutes with 500 watts dryer to simulate the passage of the night.

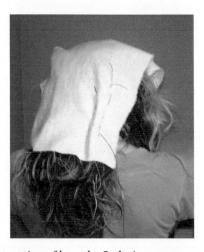

Figure 8.2 Reconstruction of how the Sudarium was sewn to the hair.

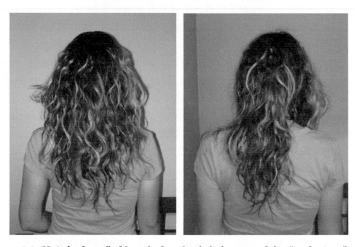

Figure 8.3 Hair before (left) and after (right) the use of the "sudarium".

The crown of thorns was simulated with pure blood from a second extraction by the nurse like the previous one. With the same syringe, once removed the needle; 12.5 cc of blood were deposited as drops at about 50 points on the scalp under the hair of the head. Hair diadems were used to simulate the fastening effect of the crown of thorns. It was placed surrounding the head at the temples and passing over the nape to hold the hair as the crown would. Then, another drying session was applied like the previous one.

To simulate the sweat and perspiration caused by the effort during the Calvary road carrying the weight of the cross, 25 cc of physiological serum were applied. Then, other 12.5 cc of blood from a third extraction were applied just like last time to simulate the reopening of the crown wounds caused by various blows and falls again, Then, another session of dryer was applied this time allowing the air to drag very soft sand from the desert placed in front of the dryer representing the dust of the road assuming some falls of Christ. The dryer blew around the head.

The diadem was then removed, assuming that the crown was removed after death and before the Sudarium was placed on it. At the end, the volunteers' hair remained covered with serum, dust, and blood (Fig. 8.4).

Figure 8.4 Volunteers for the experiment with their hair soaked with serum, dust and blood. The man wanted to remain anonymous.

After this simulation of the first stage of the Passion, we wrapped the head with a sudarium-like cloth and performed an accelerated drying with a hairdryer to simulate the course of two hours that the Sudarium remained wrapping the head.

Once the cloth was removed, we verified that the hair remained thick and compact. The tuft at the back of the neck was thicker and more compact that in the previous simulation (Fig. 8.5). We observed how the hair which was covered with serum, dust and blood hardened and it became firm when the drops of blood corresponding to the wounds of thorns dried.

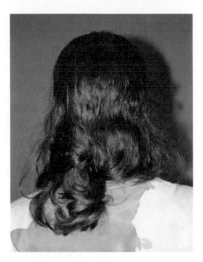

Figure 8.5 Hair tuft after simulation of the Christ Passion.

As we have seen, the analysis of the Sudarium shows that the hair was pulled back and sewn to the Sudarium on the back of the head. The experiments performed support that once the Sudarium was carefully removed from the corpse, the hair on the back would have maintained its ponytail shape.

If this Sudarium were used for the same corpse as that of the Shroud of Turin, the image of this hair bundle would be transferred to the Italian cloth soon after.

This lock of hair delimited by the use of the Sudarium could therefore be the Shroud's ponytail. It would be recorded by the image formation process. We think this is the most plausible explanation for the existence of the Man of the Shroud's hair tress. If this is true, the hair would not be washed.

This explanation is confirmed when the stains on the nape are superimposed on the nape on the image of the Shroud. When we do that as we saw in previous chapter, the stained corner of the Sudarium lies just over the angle in the ponytail on the Shroud. As it can be seen in Fig. 8.6, the hair descends obliquely from the left side of the head towards the centre of the back and just where the edge of the Sudarium would be, the hair makes a break to fall vertically. The Oviedo cloth would have held the hair along its oblique part up to the edge of the Sudarium and from this point the hair would be free to fall vertically straight. This provides further explanation of the Shroud Man's ponytail shape.

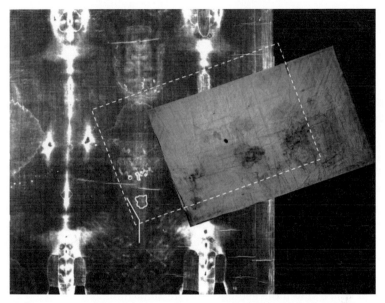

Figure 8.6 When the stains on the nape of the neck on both cloths are superimposed, the edge of the Sudarium falls on the change of direction of the ponytail on the Shroud.

An additional observation that was obtained from the experiments was regarding the distribution of dot shaped stains. Immediately after putting the cloth on the stained head, some dot shaped stains appeared which corresponded to the drops deposited on the scalp. The specific area where the first stains appeared was just the same occipital part as in the Shroud (Fig. 8.7). Sweat and blood flow from the upper part of the head falling adhered to the skin. They go from the inside to out at the occipital area where the skull starts its negative slope.

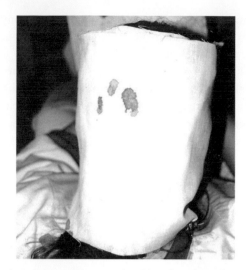

Figure 8.7 During the experiment, the first dot shaped stains appear on the same nape part as in the Shroud.

We must provide here an explanation for a difficulty that arose during the geometrical study of the Sudarium. If we measure the distance between the tip of the nose and the occipital zone, it results too short: 25 cm. This measure gives the half circumference of the head along the horizontal plane. This circumference corresponds to a too narrow head in reference to its height. In the 2nd congress of the Sudarium in 2007, Dr Jaime Izquierdo built an accurate 3D computer model of a head which suited the geometry of the Sudarium.[3]

In Fig. 8.8 we can see the digital models of the Sudarium and a head compatible with the Sudarium compared to a common head. When we fit the height, the width of the head model is too narrow. The origin of this problem is the position of the occipital dot shaped stains which is at a short distance from the position of the tip of the nose.

One possibility is to consider the matching of the dot shaped stain in the occipital area wrong and review every conclusion linked to this assumption. However, we can provide another possible

[3]Izquierdo, J., and Bernedo, J. A. (2007). *Tratamiento informático del Sudario de Oviedo.* Proceedings of the 2nd International Congress for the Sudarium of Oviedo. Oviedo. 13–15 April 2007.

solution to this inconsistency. During the experiment of the first approach with a girl, the cloth was sewn to the tress of hair.

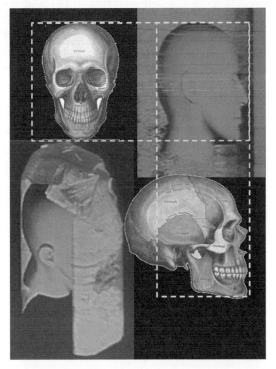

Figure 8.8 3D computer model of a head corresponding to the Sudarium (bottom left). It is too narrow when compared to a common head.

Figure 8.9 The Sudarium was tightened strongly around the face.

When it was tightened to adjust it strongly to the contour of the face (Fig. 8.9) the tress shifted from its central position to the left approaching to the left ear (Fig. 8.10) and the distance between the seam and the tip of the nose became 25 cm, the same one as in the Sudarium of Oviedo! We remark that the dot shaped stains were in the hair and not in the scalp. Once the Sudarium was removed, the hair almost regained its previous position and the stains of this hair returned to the occipital area. Soon after, the Shroud covered this part of the head and was stained with the same stains as they had stained the Sudarium. Therefore, we can maintain the simultaneous matching of the nose and the dot shaped stains with the image of the Shroud of Turin.

Figure 8.10 While the cloth is tightened, the tress shifted from its central position to the left.

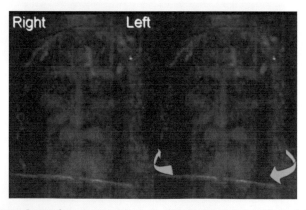

Figure 8.11 Lateral tresses of hair in the image of the Shroud indicate the displacement generated by the Sudarium.

A clue of the displacement of the lateral tresses is observed in the image of the Shroud because the left tress is ahead of the right tress (Fig. 8.11).

8.2 Hair Locks along Cheeks

There are more characteristics which support that the Sudarium of Oviedo was used on the Man on the Shroud. On the Shroud, the image of the head shows the lateral tresses of the hair[4] aligned next to the cheeks and reaching their end to the front of the shoulders. However, if a corpse is lying on its back, it would be expected that these locks would fall towards the back, leaving the side of the face and the ear free (Fig. 8.12).

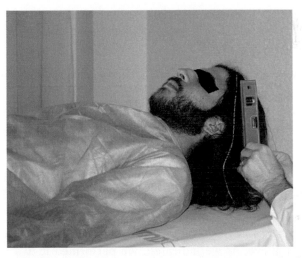

Figure 8.12 Clean hair falls vertical when the person lies on its back.

On the contrary, the 3D image reconstruction based on data from the Shroud shows the hair of the Man running along its cheeks (Fig. 8.13[5]). Some researchers have tried to explain this anomaly by supposing that the corpse had a chin guard that held

---

[4]Fanti proposes that the lateral tresses can be the typical *payots* which are worn by some Orthodox Jewish people. Fanti, G., and Malfi, P. (2020). *The Shroud of Turin. First Century after Christ!* Second Edition. Jenny Stanford Publishing Pte. Ltd. Singapore. p. 40, p. 112, and p. 129.

[5]Sculpture of the Man in the Holy Shroud created by Luigi Enzo Mattei.

the tresses of the sides.[6] In the first chapter, we analysed this hypothesis claiming to be supported by the evangelical account and identifying the chin guard as the Sudarium mentioned by Saint John.

 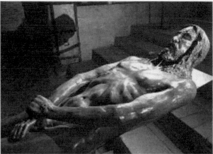

Figure 8.13 3D reconstruction based on data from the Shroud by computer (left) and Mattei sculpture (right). The tresses of hair on both sides of the head remain aligned next to the cheeks.

As we explained in the first chapter, the alleged chin band is not well justified in a corpse that already remained with its mouth closed after having leaned its head against the chest when dying. We have already dismissed the hypothesis. Another explanation could be the following:

> The use of the Sudarium of Oviedo is enough to keep the hair in the position which appears on the Shroud without the need of any other element.

To verify this fact, we performed the experimental tests described above. As we indicated, we performed it two times on two different individuals, stimulating the process that the hair of Jesus Christ underwent during his passion. The objective was to determine whether the hair was compacted enough at the end of the whole process that Jesus Christ presumably underwent during His late hours before dying. We wondered whether the use of the Sudarium justifies the fact that the Man in the Shroud has his hair along his face even when lying down.

---

[6]Rodella, S. Real Colegiata de San Isidro. Madrid. Exposición esculturas de Sergio Rodella (Italia). OSMTJ-OSMTH. October 2021.

At the beginning of the experience, when the volunteers had their hair clean, we verified how their hair fell close to the vertical when lying down (Fig. 8.12). During the first experiment with the author's daughter, it was not noticed that she had removed her hair from her face and gathered it above and behind her ear to avoid being disturbed. This gesture, however, would have been impossible in a crucified person. In the second experiment, the man was instructed not to manipulate his hair or face and, when he was asked to tilt his head, the tresses came forward, partially hiding his face. This posture corresponds to the gospel description and the one reconstructed in the sculpture promoted by Msgr Ricci (Fig. 8.14[7]).

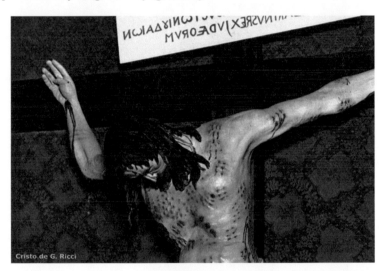

Figure 8.14 The sculpture promoted by Msgr Giulio Ricci shows the tresses falling next to the cheeks when the head is leaned against the chest.

When his head was bowed forward, the linen cloth was placed around his head, leaving his hair fixed on his cheeks (Fig. 8.15). By means of a plastic tube, with its edge of exit placed at the mouth and the incoming edge over the head, a little amount of

---

[7]Ricci, G. (1978). *The Way of the Cross in the Light of the Holy Shroud.* Center for the Study of the Passion of Christ and the Holy Shroud: Milwaukee WI. Second edition. Reprinted 1982. p. 61. According to Fanti, the arms of Jesus were more outstretched. Fanti, G., and Malfi, P. (2020). *The Shroud of Turin – First Century after Christ!* Second Edition. Jenny Stanford Publishing Pte. Ltd. Singapore. pp. 347–348.

blood was dropped close to the mouth simulating the leak of oedema observed in the Sudarium of Oviedo (Fig. 8.16).

Figure 8.15 By bowing the head, the tresses of hair are in front of the face and the sudarium fixes them in that position.

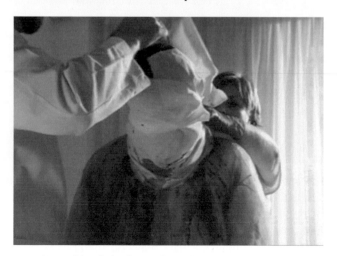

Figure 8.16 Some blood discharged at the mouth level to simulate the lower part of the main stain of the Sudarium of Oviedo.

Once the cloth was fixed around the head, another session of hair dryer was applied as we said above, emulating the approximate time that elapsed between the placement of the Sudarium on the Man still on the cross and His descent to the prone position.

With the head more inclined, another drying session was carried out, emulating the approximate time that had elapsed at the foot of the cross.

Finally, the volunteer was placed on his back and the sudarium was removed. In the case of the girl, the results show a total rigidity of the hair from the level where the diadem was upwards. From this level down, in comparison, a decreasing stiffness is manifested but that does not just disappear even at the ends.

In the case of the male, who was not allowed to move his hair, his ears, part of the cheekbones, and jaw remained covered by hair after removing the cloth. The tresses reached the front of the shoulders (Fig. 8.17).

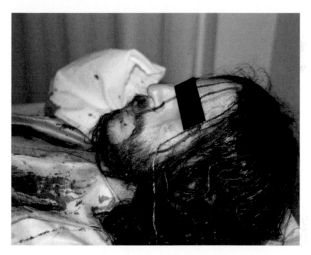

Figure 8.17 After removing the sudarium dummy, the lateral tresses covered the cheekbones and reached the shoulders.

It was verified, therefore, that the use of a sudarium such as that of Oviedo justifies itself the irregularity of the tresses of the Man of the Shroud. The hair defying the law of gravity that can be seen in the image of Turin is justified because of the fixing effect that the Sudarium causes on hair compacted by the stains of blood and dust.

Additional conclusions were drawn. Dirty hair stained with dried blood offers a sufficiently consistent volume to support the apparent seam of the Sudarium.

In summary, the placement of the tresses along the cheeks of the Man of the Shroud, despite the horizontal position of the body, is explained by the fixative effect the Sudarium had, since the hair was held in that position for over two hours of use.

We see how the information obtained from the Sudarium of Oviedo allows better interpreting the information of the Shroud of Turin (and vice-versa).

Chapter 9

Liturgy for the Sudarium in the Church

In this chapter we will describe the veneration of the Sudarium in the Cathedral of Oviedo since the opening of the Holy Ark until today. Here we follow the works of Janice Bennet[1] and Enrique López[2] mainly improved with the contributions from the two congresses held in Oviedo for the Sudarium.

9.1 Opening of the Holy Ark

Although the ceremony of opening of the Holy Ark was not precisely dedicated to the Sudarium, we present it as the first possible Liturgy related to the Sudarium of Oviedo. The veneration of the relics that arrived in the Holy Ark and the veneration of the Holy Sudarium was one and the same for the first centuries after the opening of the chest in 1075. Only since the specific exposition of the Sudarium in the 16th century, can we refer to veneration of the Sudarium properly.

---

[1]Bennett, J. (2001). *Sacred Blood. Sacred Image. The Sudarium of Oviedo*. Libri de Hispania. Colorado. pp. 48–52.
[2]López Fernández, E. (2004). *El Santo Sudario de Oviedo*. MADU. Granda-Siero. pp. 126–130.

The Sudarium of Oviedo: Signs of Jesus Christ's Death
César Barta
Copyright © 2022 Jenny Stanford Publishing Pte. Ltd.
ISBN 978-981-4968-13-3 (Hardcover), 978-1-003-27752-1 (eBook)
www.jennystanford.com

While the Ark remained closed, no one knew about the contents of the relics that were inside. However, they expected the chest to contain many sacred relics. As we revealed in Chapter 2 about the history of the Sudarium, the event was presided by the highest authorities: King Alfonso VI, his sister Lady Urraca, the *Infanta* Lady Elvira, Knight El Cid, several bishops, and the highest ecclesiastical dignitaries. Previously, they had to celebrate services and follow extraordinary fasts. To enhance the event, songs and censers of incense accompanied the occasion.

When the news of the sacred treasure reached the Christian world, popular devotion for the Holy Chamber of the Cathedral of Oviedo grew, despite the access to Oviedo being difficult because of the surrounding mountains. Soon the sanctuary became a popular pilgrimage destination. Pilgrims who went to Santiago de Compostela, the famous pilgrimage sanctuary, coined the expression: Who goes to Santiago and not to San Salvador (Oviedo Cathedral), honours the servant and forsakes the Lord[3].

The pilgrimage to the Cathedral of Oviedo and the associated indulgences are testified by the sheets of paper (*buletas*) delivered to the pilgrims in exchange of offerings. In 1488 the sheets were written by hand but from 1493 onwards, they were impressed, always by thousands. There were written in several languages, such as Spanish, Latin, and French.

As we have already noted in a previous chapter, in these sheets the list of relics and the indulgences granted to the pilgrims was described. In the first *buletas*, the list of arrived relics inside the Holy Ark, mentioned the shroud (*sindone*) but did not include the Sudarium. Since 1576, the Sudarium has been included in the list of relics just after the shroud.

> ... *magnam partem de Sindone Domini, qua involutus iacuit in monumento. Eiusque pretiosum Sudarium suo cruore sanctissimo intinctum, quo speciosissima eius facies cum sacratissimo capite cooperta atque involuta fuit quodque ter singulis annis ea qua decet reverentia ostenditur. Videlicet in festo sollemni dictarum sanctarum reliquiarum, quod incidit die decima tertia Martii,*

---

[3]Bennett, J. (2001). *Sacred Blood. Sacred Image. The Sudarium of Oviedo*. Libri de Hispania. Colorado. p. 45.

ac feria sexta in Parasceve. Necnon in festo Exaltationis sanctae Crucis die 14 Septembris et in diebus festivis contentis intra tempus Iubilei quoid infra declarat[4]

[... A large part of the shroud of the Lord, in which he laid wrapped in the tomb. His precious Sudarium stained with his most holy blood. With it his most beautiful face and his most sacred head were covered and wrapped. It is shown, with due reverence three times every year, namely, on the solemn feast of the aforementioned holy relics, which happens on the thirteenth of March, on Good Friday and also on the feast of the Exaltation of the Holy Cross on 14 September and on holidays included within the time of the Jubilee.]

9.2   Yearly Exhibitions and Indulgences

With the increase in the number of pilgrims to venerate the Holy Ark, the Chapter members petitioned Pope for a Jubilee. In the papal bull of 10 November 1438, a plenary indulgence was conceded to those who visit the Cathedral of Oviedo on the day of the Exaltation of the Holy Cross or during the eight days before or after, during the year in which such feast would occur on Friday, and for those who would present offerings for the building fund of the Cathedral. Pope Pius IV, on 13 November 1561, extended the indulgences magnanimously[5].

According to the sheets, the following indulgences are granted to pilgrims[6].

1493 and 1535: remission of a third of the penalty imposed for sins, partial indulgences of 1004 years and 6 and a half quarantines, and plenary indulgence of the Jubilee gained on the day of the Exaltation of the Holy Cross. This privilege was extended by Sixtus IV in 1480 to the fifteen days before or after in the years in which the said fest falls on Friday.

---

[4]López Fernández, E. (2004). *Las Reliquias de San Salvador de Oviedo*. MADU. Granda-Siero. p. 238.

[5]Bennett, J. (2001). *Sacred Blood. Sacred Image. The Sudarium of Oviedo*. Libri de Hispania. Colorado. p. 51.

[6]López Fernández, E. (2007). *La veneración del Sudario de Oviedo a través de la historia*. Proceedings of Oviedo relicario de la Humanidad. Actas del II Congreso sobre el Sudario de Oviedo. Oviedo. p. 360.

Pilgrims came from several countries; those mentioned include Germany, England, France, and Holland.

As we see it, unexpectedly, the Sudarium (which is not "a part" as it is said for the shroud) took prominence and reached the second place among the list of relics. Moreover, it started to have its own exhibition three times a year. In 1557, and presumably years before, the Sudarium was shown from the balcony of the Holy Chamber (Fig. 9.1).

Figure 9.1 Balcony from where the Sudarium was exhibited in 16th century.

1561: Pius IV extended the period of the privileges from fifteen to thirty days in the years in which the fest fell on Friday.

A detailed description of the veneration is provided by Ambrosio de Morales on 1572[7]. According to him, the three days of exhibition in a year were the Good Friday and the two feasts of the Holy Cross (3 May and 14 September). To grant him to fulfil an accurate account of the veneration of the Sudarium an extraordinary exposition was allowed on the feast of St James. However, the event was announced throughout the city as it happened on the official days. The church was embellished with rich cloths and sliding black velvet curtains in the path.

---

[7]Florez, H. (1765). *Viage de Ambrosio de Morales*. Madrid. Antonio Marín. p. 79.

The Bishop celebrated Pontifical Mass as he was doing for these exhibitions and preached asking for appropriate devotion for the Sacred Relic. On finishing the mass, he and his assistants went to the Holy Chamber and took out the Holy Sudarium covered with his veil. They walked to the balcony singing prayers and when they arrived, they opened the curtains of the balcony, each one running to its side. With the Sudarium in the middle, the Bishop removed the taffeta veil to show the Sudarium and people below started to sing the *Miserere*. Once the exhibition was finished, the Bishop had to return to the Holy Chamber. Before the relic was covered again, those who were with him could see the Holy Sudarium very closely. The Bishop with those who accompanied him sang the *Miserere* until the sacred relic was put in its box.

This practice lasted until recently. In 1845, the English travel writer, Richard Ford, wrote an interesting account of his visit to the cathedral of Oviedo[8]. He referred to the visit of Morales and remarked that the ecclesiastic trembled before these gold enshrined objects with fear and reverence. Ford also says that the Sudarium was exhibited publicly three times a year, and after the bishop had preached, it was displayed from a balcony. At this date, the peasants still held up loaves, beads, and other objects, in the belief that they will be provided with a nutritious and medicinal quality. Today, no one holds up bread or any another object.

Between 1967 and 1982 several changes issued from Rome reordered the discipline concerning indulgences. Finally, plenary indulgence was granted on the feast of the Exaltation of the Holy Cross on 14 September and on 21 September, the feast of St Matthew, and during the eight days from 14 to 21 of September to those who visited the Cathedral Church of Oviedo and received the sacraments of Reconciliation and Eucharist. The two dates in September are the first and last days of the Jubilee of the Holy Cross.

The conditions to receive the plenary indulgence are:

– A pious visit to the Cathedral of' Oviedo and praying to Our Father and Creed,

---

[8]Bennett, J. (2001). *Sacred Blood. Sacred Image. The Sudarium of Oviedo*. Libri de Hispania. Colorado. p. 49.

 – The sacrament of Reconciliation,
 – Receiving the Eucharist, and
 – Prayer for the intentions of the Holy Father (Our Father and Hail Mary, or another pious prayer)[9].

Some years after the Second Vatican Council had once fixed the guidelines by the highest authorities, some modifications were made to the traditional practice[10]. Thus, on 20 May 1982, the canons' Chapter agreed that the blessing with the Holy Sudarium should also be given at the end of the Via Cruces, on the afternoon of Good Friday, a few hours after the celebration of the divine services. This public benediction was not imparted from the balcony, but from the main altar itself.

Two years later, on 3 March 1984, the Chapter of canons also agreed that the Holy Sudarium would preside from the main altar over the entire celebration of the Via Cruces on Good Friday and not be limited only to the final blessing.

In the same year, on 22 September, it was decided to change the hour of the Mass where the blessing with the Holy Sudarium was given. The public benediction was passed from the 9:30 mass to the 12:00 mass.

Pope John Paul II visited the Holy Chamber in 1989. He venerated the relics in the Holy Ark, the Cross of the Angels, and the Holy Sudarium.

For centuries, the Sudarium was displayed by the bishop holding the frame by two handles placed in the longest sides of it. The bishop blessed the whole world directing it towards each of the cardinal directions (Fig. 9.2).

Nevertheless, in April 2014 the conservation system of the Sudarium changed. It is now preserved in a sealed urn in accordance with the EDICES recommendations. This urn or reliquary has the upper part made of transparent safety glass. The primitive frame of silver has been kept just for a decorative purpose (Fig. 9.3). It is held by two young men when it must be

---

[9]Bennett, J. (2001). *Sacred Blood. Sacred Image. The Sudarium of Oviedo*. Libri de Hispania. Colorado. p. 51.

[10]López Fernández, E. (2004). *El Santo Sudario de Oviedo*. MADU. Granda-Siero. p. 129.

moved (Fig 9.4). For the exhibition, the reliquary is taken out from its large container where it is preserved the rest of the time (Fig. 9.5).

Figure 9.2 Sudarium exhibition until 2014. The bishop holds the frame by the longest sides (© CES).

Figure 9.3 New sealed reliquary while the exhibition.

During the main mass at 12:00 the Sudarium presides over the celebration on a support that holds the 18 kg of the set (Fig. 9.6). At the end of the mass, the worshippers can approach to venerate it.

Figure 9.4 New reliquary being carried out for the exhibition.

Figure 9.5 The author in front of the Sudarium inside its container instants before it is to be translated to the main altar of the Cathedral on 21 September 2020, St Mathew fest.

Apart from the veneration events, the visitor is allowed to access to the Holy Chamber always in the company of a person who guides in the cathedral. People can see through a grille the container of the Sudarium with a pretty facsimile that covers the transparent side of the container and protect the Sudarium from the environment light. The container is located in the middle and around it, other relics and objects are placed (Fig. 9.7).

On exceptional occasions, the cult of the Holy Sudarium was not organized inside the Cathedral. In Spain there is a strong tradition to take out the images of the saints in procession through the streets on festive days. On a few occasions, the Sudarium walked the streets of Oviedo as well for veneration. More properly, these events were associated with supplication, invoking with it the intercession of the Lord to solve a major problem, such as plague, the French invasion, or storms.

Figure 9.6 Nowadays, the Sudarium presides over the main Mass at 12:00 on a support that holds the 18 kg of the reliquary.

Such was the case on 15 September 1598 when the Sudarium was taken only up to the Cloister and under the Towers due to an important plague.

In September 1638 there was a proposal to take the Sudarium outside because the war against France was still in progress. At least, an exposition was prepared organizing shifts as if it were the Eucharist for the feast of Saint Matthew on 21 September.

In April 1692 and July 1697 as a compliment to the prayers, the Sudarium went outside the Cathedral as well, just to the door, without making a procession as a remedy against storms.

In May 1808, because of the menacing war against the French and Napoleon, the Chapter of canons with the agreement of the bishop allowed the procession of the Sudarium along several streets of Oviedo. The Sudarium, as the most venerated relic, accompanied Our Lady and the Cross of the Victory. They prayed to God to preserve their holy religion and the King Ferdinand the Seventh against the invasion of the French. The Bishop carried the Sudarium throughout the procession flanked by files of royal guards while the multitude admired the relic.

Figure 9.7 Relics of the Oviedo Cathedral behind the grille in the Holy Chamber.

Once the reconstruction of the Holy Chamber, blown up in October 1934 was completed, on 5 September 1942 the Holy Relics were incorporated into it[11]. A procession was organized to recover their dignity and was documented by photography (Fig. 9.8).

A long record of private exposures to important people is provided by Enrique López[12]. However, the respect for the Sudarium preserved by the owners in the Cathedral prevents any

---

[11]López Fernández, E. (2004). *El Santo Sudario de Oviedo*. MADU. Granda-Siero. p. 130. López Fernández, Enrique. (2007). *La veneración del Sudario de Oviedo a través de la historia*. Proceedings of Oviedo relicario de la Humanidad. Actas del II Congreso sobre el Sudario de Oviedo. Oviedo. pp. 362–363.

[12]López Fernández, E. (2007). *La veneración del Sudario de Oviedo a través de la historia*. Proceedings of Oviedo relicario de la Humanidad. Actas del II Congreso sobre el Sudario de Oviedo. Oviedo. pp. 363–364.

approach by simple curiosity. For them, the Sudarium is always the Holy or the Most Holy Sudarium and only those in cases for veneration do they grant permission for the visit.

Figure 9.8 Procession with the Sudarium along the streets of Oviedo probably in the 20th century (© CES).

9.3   Specific Mass for the Holy Sudarium

It is stated in the documentation of the Cathedral of Oviedo that since ancient times a Holy Votive Mass was celebrated in honour of the Holy Sudarium which was shown annually to the worshippers as an authentic relic of the Lord's Passion. Such Mass stopped being celebrated after the dispositions of Pope Urban VIII promulgated between 1625 and 1642[13]. This Pope tried to introduce defined regulations in the liturgical and venerations of saints and relics policy that were considered a kind of anarchy and disorder. It was the reaction to the Protestant Reformation that accused Catholics of promoting events that were not always adequately serious and theologically founded.

---

[13]González, F. (2007). *Liturgia de la Misa y de las Horas del Santo Sudario propia para la Catedral y Archidiócesis de Oviedo.* Proceedings of Oviedo relicario de la Humanidad. Actas del II Congreso sobre el Sudario de Oviedo. Oviedo. pp. 311–353.

Nevertheless, the bishop Antonio Valdés Herrera (1636–1641) requested permission for recovering the Mass in honour of the Holy Sudarium and presented a text of the Mass to be approved. However, at this first instance, the Congregation of Rites answered negatively on 14 January 1640 to the request.

After several centuries, in December 2006, the bishop of Oviedo, D. Carlos Osoro presented before the Sacred Congregation for Divine Worship a new request for the Votive Mass and for the Liturgy of the Hours as well in honour of the Holy Sudarium. Among the reasons to such a request, he presented the results of the scientific research carried out on the Holy Cloth. He mentioned the 1st Congress for the Sudarium of Oviedo whose proceedings had been sent to the Congregation. Moreover, he reminded to the institution that this type of Votive Mass is already being conducted for the Shroud of Turin.

The positive response was received on 17 January 2007. A final version of the Votive Mass and of the Liturgy of the Hours was approved by the Supreme Authority of the Congregation. The official text, of course, in Spanish, is provided in annex. Perhaps, some reader of this book will have the opportunity of following the Mass one of the days of the Sudarium exhibition with the help of this annex.

The collect prayer is addressed to the Father asking him for the grace to those who venerate His image in the Holy Sudarium to contemplate in heaven the infinite beauty of the face of His Son. Jesus Christ.

In the prayer over the offerings, the argument continues imploring the Father to set His eyes on the suffering face of Christ His Son and grant the worshipers to become similar to Him.

The preface is the same as the Liturgy of the Sacred Heart of Jesus remembering He was raised on the cross and from His wound in his side water and blood flowed.

The proposed scripture readings are from the New Testament. The first refers to Peter's announcement of the Mystery of Jesus, in front of both the Roman centurion Cornelius and the members of his family. And the Gospel is that from John (20: 1-9) that tells the resurrection of Christ and spreads the news of the existence of the "Sudarium with which had covered His head".

The second reading for the office of the Liturgy of the Hours, chooses the commentary on the Psalms of Saint Ambrose (PL 14, 1185), dealing with the face of God which enlighten us with the light of it.

The Liturgy of the Holy Sudarium granted by the Apostolic See to the Archdiocese of Oviedo is a catechesis on the Paschal Mystery.

9.4 Annex: Transcribed Text of the Votive Mass and Liturgy of the Hours (Spanish)

MISA VOTIVA[14]

En los días del tiempo ordinario en que se permite la celebración de un oficio votivo, se puede celebrar el oficio votivo del Santo Sudario.
Esta misa se dice con vestiduras de color rojo.

ANTÍFONA DE ENTRADA   Jn 19:40

Tomaron el cuerpo de Jesús y Io vendaron todo, con los aromas, según se acostumbra a enterrar entre los judíos.

ORACIÓN COLECTA

Oh, Padre, que en la pasión has glorificado a tu Hijo Jesucristo y Io has constituido Señor, en la resurrección de entre los muertos, concédenos, a los que veneramos su imagen en el santo Sudario, poder contemplar en el cielo la hermosura infinita de su rostro. Él, que vive y reina contigo.

ORACIÓN SOBRE LAS OFRENDAS

Acoge, Padre, estos dones y plegarias, puestos tus ojos en el rostro doliente de Cristo, tu Hijo, y concédenos hacernos semejantes a él, que se ofreció a sí mismo como víctima inmaculada. Él, que vive y reina por los siglos de los siglos.

---

[14]Texts taken from González, F. (2007). *Liturgia de la Misa y de las Horas del Santo Sudario propia para la Catedral y Archidiócesis de Oviedo.* Proceedings of Oviedo relicario de la Humanidad. Actas del II Congreso sobre el Sudario de Oviedo. Oviedo. pp. 346–353.

PREFACIO

EL INMENSO AMOR DE CRISTO

V/. El Señor esté con vosotros.
R/. y con tu espíritu.
V/. Levantemos el corazón.
R/. Lo tenemos levantado hacia el Señor.
V/. Demos gracias al Señor, nuestro Dios.
R/. Es justo y necesario.

En verdad es justo y necesario, es nuestro deber y salvación darte gracias siempre y en todo lugar, Señor, Padre santo, Dios todopoderoso y eterno, por Cristo, Señor nuestro.

El cual, con amor admirable se entregó por nosotros, y elevado sobre la cruz hizo que de la herida de su costado brotaran, con el agua y la sangre, lo sacramentos de la Iglesia; para que así, acercándose al Corazón abierto del Salvador, todos puedan beber con gozo de la fuente de la salvación.

Por eso, con los ángeles y arcángeles y con todos los coros celestiales, cantamos sin cesar el himno de tu gloria:
Santo, Santo, Santo ...
También puede utilizarse el prefacio de la exaltación de la santa Cruz o alguno de lo. prefacios comunes.

ANTÍFONA DE COMUNIÓN   Jn 20:6,8

Llegó Simón Pedro y entró en el sepulcro: vio el sudario con que le habían cubierto la cabeza, enrollado en un sitio aparte.

ORACIÓN DESPUÉS DE LA COMUNIÓN

Después de participar del sacramento de tu amor, te pedimos, Dios nuestro, la gracia de parecernos a Cristo en la tierra para merecer compartir su gloria en el cielo.

Él, que vive y reina por los siglos de los siglos.

LECTURAS

En los días del tiempo ordinario en que se permite la celebración de un oficio votivo, se puede celebrar el oficio votivo del Santo Sudario.

PRIMERA LECTURA

Los que creen en él reciben el perdón de los pecados.

LECTURA DEL LIBRO DE LOS HECHOS DE LOS APÓSTOLES
10: 34a. 37–43

En aquellos días, Pedro tomó la palabra y dijo:

«Conocéis lo que sucedió en el país de los judíos, cuando Juan predicaba el bautismo, aunque la cosa empezó en Galilea. Me refiero a Jesús de Nazaret, ungido por Dios con la fuerza del Espíritu Santo, que pasó haciendo el bien y curando a los oprimidos por el diablo, porque Dios estaba con él.

Nosotros somos testigos de todo lo que hizo en Judea y en Jerusalén. Lo mataron colgándolo de un madero. Pero Dios lo resucitó al tercer día y nos lo hizo ver, no a todo el pueblo, sino a los testigos que él había designado: a nosotros, que hemos comido y bebido con él después de su resurrección.

Nos encargó predicar al pueblo, dando solemne testimonio de que Dios lo ha nombrado juez de vivos y muertos. El testimonio de los profetas es unánime: que los que creen en él reciben, por su nombre, el perdón de los pecados.»

Palabra de Dios.

SALMO RESPONSORIAL   Sal 117: 1–2.16ab–17. 22–23 (R.: 1)

R. Dad gracias al Señor porque es bueno, porque es eterna su misericordia.

Dad gracias al Señor porque es bueno, porque es eterna su misericordia. Diga la casa de Israel: eterna es su misericordia. R.

La diestra del Señor es poderosa, la diestra del Señor es excelsa. No he de morir, viviré para contar las hazañas del Señor. R.

La piedra que desecharon los arquitectos es ahora la piedra angular. Es el Señor quien lo ha hecho, ha sido un milagro patente. R.

ALELUYA   Jn 20:1

El primer día de la semana, María Magdalena fue al sepulcro al amanecer, cuando aún estaba oscuro, y vio la losa quitada del sepulcro.

EVANGELIO

Vio y creyó.

LECIURA DEL SANTO EVANGELIO SEGÚN SAN JUAN 20: 1–9

El primer día de la semana, María Magdalena fue al sepulcro al amanecer, cuando aún estaba oscuro, y vio la losa quitada del sepulcro.

Echó a correr y fue donde estaba Simón Pedro y el otro discípulo, a quien tanto quería Jesús, y les dijo:

-«Se han llevado del sepulcro al Señor y no sabemos dónde lo han puesto.» Salieron Pedro y el otro discípulo camino del sepulcro. Los dos corrían juntos, pero el otro discípulo corría más que Pedro; se adelantó y llegó primero al sepulcro; y, asomándose, vio las vendas en el suelo; pero no entró.

Llegó también Simón Pedro detrás de él y entró en el sepulcro: vio las vendas en el suelo y el Sudario con que le habían cubierto la cabeza, no por el suelo con las vendas, sino enrollado en un sitio aparte.

Entonces entró también el otro discípulo, el que había llegado primero al sepulcro; vio y creyó.

Pues hasta entonces no habían entendido la Escritura: que él había de resucitar de entre los muertos.

Palabra del Señor.

LITURGIA DE LAS HORAS

En los días del tiempo ordinario en que se permite la celebración de un oficio votivo, se puede celebrar el oficio votivo del Santo Sudario.

INVITATORIO

Ant. A Cristo, el Señor, que en el Santo Sudario nos dejó un signo de su redención, venid adorémosle.

El salmo invitatorio como en el Ordinario.

OFICIO DE LECTURA

HIMNO

PRIMERA LECTURA

De la feria correspondiente o de la solemnidad del sagrado Corazón de Jesús (LH III, p. 540).

SEGUNDA LECTURA

Del «Comentario a los salmos» de san Ambrosio, obispo (PL 14, 1185)

Brille sobre nosotros la luz de tu rostro, Señor.

Pensamos que Dios oculta su rostro cuando nos encontramos en cualquier tribulación. Entonces se extiende un velo tenebroso sobre nuestro espíritu que nos impide percibir el fulgor de la verdad. Pero si Dios se interesa por nosotros y se digna visitarnos, estamos seguros de que nada puede ocurrirnos en la oscuridad. El rostro del hombre es como una luz para quien lo mira. Por él alcanzamos a reconocer a los desconocidos o reconocemos a una persona que recordamos. Al mostrar el rostro somos identificados. Por tanto, si el rostro del hombre es como una luz, ¿cuánto más no lo será el rostro de Dios para quien lo contempla? El Dios que dijo: Brille la luz del seno de las tinieblas, le ha encendido en nuestros corazones, haciendo resplandecer el conocimiento de la gloria de Dios, reflejada en el rostro del Mesías.

Hemos escuchado, pues, cómo Cristo brilla en nosotros. Él es, en efecto, el resplandor eterno de las almas, enviado por el Padre a la tierra para iluminarnos con la luz de su rostro, para que podamos ver las cosas eternas y celestes, nosotros, que antes estábamos inmersos en las tinieblas de la tierra.

Pero ¿por qué hablar de Cristo, cuando también el apóstol Pedro dijo a aquel lisiado de nacimiento: «Míranos»? Él miró hacia Pedro y fue iluminado por la gracia de la fe; en efecto, no hubiera recibido el don de la salud si no hubiese creído por la fe.

Sin embargo, pese a toda esta luz de la gloria presente en los apóstoles, Zaqueo prefirió la de Cristo. Oyendo que pasaba el Señor, subió a un árbol, porque al ser bajo de estatura, no podía verle con tanta multitud. Vio a Cristo y encontró la luz; lo vio y, de robar anteriormente las cosas de los demás, pasó a distribuir las suyas propias... «¿Por qué me escondes tu rostro?». O mejor: aunque se desvíe la mirada de nosotros, permanece igualmente en nosotros la impronta de tu rostro. La tenemos en nuestros corazones y resplandece en lo más íntimo de nuestro espíritu: nadie puede subsistir, si tú apartas completamente de nosotros tu rostro.

RESPONSORIO　　　　Jn 15:1.5.9

R/. Yo soy la verdadera vid, vosotros los sarmientos; * El que permanece en mí y yo en él, ese da fruto abundante.

V/. Como el Padre me ha amado, así os he amado yo; permaneced en mi amor. * El que permanece.

La oración como en Laudes.

LAUDES

HIMNO

Las antífonas y los salmos, del día correspondiente.

LECTURA BREVE　　　　Is 63: 1–3

¿Quién es ése que viene de Edom, de Bosra, con las ropas enrojecidas? ¿Quién es ése, vestido de gala, que avanza lleno de fuerza? Yo, que sentencio con justicia y soy poderoso para salvar. ¿Por qué están rojos tus vestidos, y la túnica como quien pisa en el lagar? Yo solo he pisado el lagar, y de los otros pueblos nadie me ayudaba. Los pisé con mi cólera, lo estrujé con mi furor; su sangre salpicó mis vestidos y me manché toda la ropa.

RESPONSORIO BREVE

R/. Ten piedad de mí, Señor. * No me escondas tu rostro. Ten piedad.

V/. Tu rostro, Señor, buscaré yo. * No me escondas tu rostro. Gloria al Padre. Ten piedad.

Benedictus, ant. El primer día de la semana, María Magdalena fue al sepulcro al amanecer, cuando aún estaba oscuro, y vio la losa quitada del sepulcro.

PRECES

Adoremos a nuestro Redentor que por nosotros y por todos los hombres quiso morir y ser sepultado para resucitar de entre Io muertos, y supliquémosle, diciendo:

Señor, ten piedad de nosotros.

Señor Jesús, que como grano de trigo caíste en la tierra para morir y dar con ello fruto abundante,

- haz, que también nosotros sepamos morir al pecado y vivir para Dios.

Oh, Pastor de la Iglesia, que quisiste ocultarte en el sepulcro para dar la vida a los hombres,

- haz que nosotros sepamos también vivir escondidos contigo en Dios.

Nuevo Adán, que quisiste bajar al reino de la muerte para librar a los justos que, desde el origen del mundo, estaban sepultados allí,

- haz que todos los hombres, muertos al pecado, escuchen tu voz y vivan.

Cristo Hijo del Dios vivo, que has querido que por el bautismo fuéramos sepultados contigo en la muerte,

- haz que, siguiéndote a ti, caminemos también nosotros en una vida nueva.

Padre nuestro.

Oración

Oh, Padre, que en la pasión has glorificado a tu Hijo Jesucristo y lo has constituido Señor, en la resurrección de entre los muertos, concédenos, a los que veneramos su imagen en el santo Sudario, poder contemplar en el cielo la hermosura infinita de su rostro. Él, que vive y reina contigo.

HORA INTERMEDIA

Las antífonas y los salmos, del día correspondiente.

TERCIA

LECTURA BREVE   **Is 53: 2–3a**

Creció en su presencia como brote, como raíz en tierra árida, sin figura, sin belleza. Lo vimos sin aspecto atrayente, despreciado y evitado por los hombres.

V/. Te adoramos, oh, Cristo, y te bendecimos. R/. Porque por tu santa cruz, redimiste al mundo.

SEXTA

LECTURA BREVE   **Is 53: 4–5**

El soportó nuestros sufrimientos y aguantó nuestros dolores; nosotros lo estimamos leproso, herido de Dios y humillado; pero él fue traspasado por nuestras rebeliones, triturado por nuestros crímenes. Nuestro castigo saludable cayó sobre él, sus cicatrices nos curaron.

V/. El Señor da la muerte y la vida.

R/. Hunde en el abismo y saca de él.

NONA

LECTURA BREVE   **Is 53: 6–7**

Todos errábamos como ovejas, cada una siguiendo su camino, y el Señor cargó sobre él todos nuestros crímenes. Maltratado, voluntariamente se humillaba y no abría la boca como cordero llevado al matadero, como oveja ante el esquilador, enmudecía y no abría la boca.

V/. Me confina a las tinieblas.

R/. Como a los muertos ya olvidados.

La oración como en Laudes.

VÍSPERAS
HIMNO

Las antífonas y los salmos, del día correspondiente.

LECTURA BREVE   **1 P 1: 18–21**

Ya sabéis con qué os rescataron de ese proceder inútil recibido de vuestros padres: no con bienes efímeros, con oro o plata, sino a precio de la sangre de Cristo, el Cordero sin defecto ni mancha, previsto antes de la creación del mundo y manifestado al final de los tiempos por vuestro bien. Por Cristo vosotros creéis en Dios, que lo resucitó de entre los muertos y le dio gloria, y así habéis puesto en Dios vuestra fe y vuestra esperanza.

RESPONSORIO BREVE

R/. Te adoramos, oh, Cristo, * y te bendecimos Te adoramos.

V/. Porque con tu cruz has redimido el mundo. * Y te bendecimos. Gloria al Padre. Te adoramos.

Magníficat, ant. Pedro entró en el sepulcro, vio las vendas en el suelo y el sudario con que le habían cubierto la cabeza, no por el suelo con las vendas, sino enrollado en un sitio aparte.

PRECES

Adoremos al Salvador de los hombres, que, muriendo, destruyó nuestra muerte y, resucitando, restauró la vida, y digámosle humildemente:

Santifica, Señor, al pueblo que redimiste con tu sangre.

Redentor nuestro, concédenos que, por la penitencia, nos unamos más plenamente a tu pasión,

- para que consigamos la gloria de la resurrección.

Concédenos la protección de tu Madre, consuelo de los afligidos,

- para que podamos confortar a los que están atribulados, mediante el consuelo con que tú nos confortas.

Haz que tus fieles participen en tu pasión mediante los sufrimientos de su vida,

- para que se manifiesten en ellos los frutos de tu salvación.

Tú que te humillaste, haciéndote obediente hasta la muerte, y una muerte de cruz, santifica, Señor, al pueblo que redimiste con tu sangre.

El resto como en los Laudes.

Chapter 10

Conclusion and Future of the Sudarium

The best understood features found by the research carried out on the Sudarium have been developed in the previous chapters. In this chapter we will summarize them, and we will also recapture the unsolved questions that we identified along those chapters. We will also show the foreseen projection of the Sudarium in the scientific as well as religious field.

The Sudarium of Oviedo is a purported relic of Christ. If it were authentic, the newest and best supported conclusion from the scientific research of the EDICES and other teams would be that it was used when the Lord was already dead and still on the cross and remained around his head until his placement in the sepulchre before he was covered with his burial Shroud.

10.1   What We Know about the Sudarium

10.1.1   History

Among the relics attributed to Jesus Christ, one is the Sudarium kept in the Oviedo Cathedral. It has played an important role in the yearly events in this church since the 16th century. It is exhibited three times a year with maximum solemnity from the major altar and one of the three days of exhibition is Good

The Sudarium of Oviedo: Signs of Jesus Christ's Death
César Barta
Copyright © 2022 Jenny Stanford Publishing Pte. Ltd.
ISBN 978-981-4968-13-3 (Hardcover), 978-1-003-27752-1 (eBook)
www.jennystanford.com

Friday, which is the most important day in the Christian liturgy after the Easter Sunday. The rest of the year it is kept in the *Cámara Santa* (Holy Chamber) of the Cathedral. In the same chamber, the Holy Ark is kept since the erection of the temple. Inside this chest, many relics came to the city about the early 9th century. The Sudarium could have arrived inside this Ark. Among the relics preserved in the Holy Chamber there is also a part of a "shroud" that seems to have also arrived inside the Holy Ark. However, research has confirmed that it is not a part of the Shroud of Turin.

This Sudarium is claimed to be the one mentioned in the Gospel of John 20:6–7. When Simon Peter arrived after him [John], he went into the tomb and saw the burial cloths there, and the **cloth** that had covered his head, not with the burial cloths but rolled up in a separate place. This cloth of the Gospel could be the Sudarium of Oviedo.

To analyse the documents referring to the Sudarium of the Gospel, we need to consider that in the ancient times and before the Sudarium of Oviedo came to be known worldwide at the end of the 20th century, there were authors who identified the Sudarium of the Gospel of John not with a "handkerchief" but with the Shroud (sindone) of the Gospel, i.e., a cloth covering the entire body. It is important to bear this in mind because we do not know with certainty of what a text refers to when it says "sudarium" in Latin or its equivalent term in modern languages. It can be the big Shroud or a smaller sudarium as the one present in Oviedo.

A text from the 5th century remarks the singularity of the word *sudarium*. It was written by Nonnos of Panopolis[1], paraphrasing the Gospel of John, Chapter 20. The text says: They both together recognized the linens and the cloth (ζωστηρα) that covered his head, with a knot toward the upper back of the part that covered the hair. In the native language of Syria, it is called sudarium. He had to explain what the Sudarium was, and he added that the cloth, which he believed was the Sudarium of Jesus, had a knot. The knowledge that Nonnos had must have come from his journeys across the Holy Land. A knot left its

---

[1]Nonnus of Panopolis (Egypt) writer in 400–479 A.D. Bennett, Janice. (2001). *Sacred Blood. Sacred Image. The Sudarium of Oviedo*. (Libri de Hispania). Colorado. p. 22.

traces as well in the Sudarium of Oviedo in the same position. The Sudarium which was believed to be that of the Gospel preserved in or near the Holy Land in the 5th century had a knot like the one in the Sudarium of Oviedo. This clue leads us to identify both Sudariums[2].

A chronicle made about the years 560–570 by an anonymous pilgrim from Piacenza, Italy, supports the assumption that the Sudarium was preserved in the Holy Land at that time. The Sudarium that was on the Lord's head was in a cave on the banks of the River Jordan, worshipped by Christians before the invasion of Chosroes II (year 614)[3].

Nowadays, reliable studies on the size of sudariums in antiquity exclude identification with a shroud covering the entire body. However, there are many references in the past to a sudarium that was big enough to cover the entire body. It is the mistaken use of the word "sudarium" to refer to the Shroud, for example, in the bishop Arculf reference[4]. Another case is that of the relics whose achievement is attributed to Charlemagne. In Kornelimünster church there is the shroud in which His body was wrapped and the Facecloth that was wrapped around His head, which came originally from the treasures of Charlemagne. In this case, there is no scope for mistaking the term: one of the two cloths must be the sudarium. However, when we get enough details, we realize that none of them is a sudarium. The so-called sudarium in Kornelimünster has a size of 4 × 6 m. Astonishingly, the other cloth pretending to be the shroud is smaller, but big enough to be discarded being a sudarium. It is 1.80 × 1.05 m. A sudarium of Christ preserved in the cathedral of Mainz[5] is also a part of the "sudarium" of Kornelimünster[6] and, therefore, it could not be a true sudarium.

Among the candidates to be the Sudarium of the Gospel, the only one that stands with evidence in favour is the Sudarium of Oviedo.

---

[2]Chapter 2.
[3]Chapter 2.
[4]See below.
[5]Brodehl, T. (2009). *Die heilige Bilhildis und das Schweißtuch Christi zu Mainz*. In: Johannes A. Wolf (Hrsg.): Der schmale Pfad. Orthodoxe Quellen und Zeugnisse. Band 29. S. pp. 94–115.
[6]Matheus, M. (1999). *Pilger und Wallfahrtsstätten in Mittelalter und Neuzeit*. Franz Steiner Verlag Stuttgart. p. 107.

The Sudarium could have come in the Holy Ark whose content was listed in 1075. There is a well-documented reference for the history of the Holy Ark. However, to allow considering the ancient history of the Sudarium of Oviedo the same as that of the Holy Ark it is necessary to support the presence of the Sudarium among the relics inside the Ark. This is a subject to be reviewed. For the time being, we must differentiate the history of the Ark from the history of the Sudarium. It is only probable that both are the same.

The Holy Ark arrived in Oviedo from Toledo and documents allow us to reconstruct a path from Jerusalem to Toledo through the north of Africa and southern Spain. When Chosroes II, the king of the Persians, captured the city in 614, the Ark with the relics left Jerusalem and reached Spain. Later when the Arabs invaded Spain in 711, the Holy Ark left Toledo and headed for the North to end up in Oviedo. With the Ark already in Oviedo, popular devotion grew, and the Holy Chamber became a popular pilgrimage destination.

The list of relics that came inside the Holy Ark is included in a fully contemporaneous document on the event. It is known as "*Donación de Langreo*". In the ninth place on the list and among the seven relics of Christ is indicated "*de sudario eius*" that means "a part" of the Sudarium. The "*Donación*" does not mention any other shroud.

The chronologically following document that provides the list of relics is the inscription of the cover of the Holy Ark. It is like the list of the donation of Langreo. The Sudarium stands in the fifth place, and it does not mention a separate shroud either.

The Book of Testaments written by the bishop Pelayo mentions the two burial cloths among the relics: "a part of the Lord's shroud" (*de sindone Domini*) and "a part of the Lord's sudarium" (*de sudario Domini*). This is the major justification to think that the part of the shroud and the Sudarium arrived together in the Ark as two different cloths.

After many years of full discretion, the Sudarium came into prominence in the middle of the 16th century[7]. Since then, it has

---

[7]López Fernández, E. (2016). Historical sources about the relics of Oviedo's cathedral. Territorio, Sociedad y Poder, 11. p. 15. The minutes of 17 September 1557 are in the Archives of the Cathedral, *Acuerdos capitulares*, Libro 8, f. 586v. "*Cometieron a los ss. [] para que vean cómo se ha de hazer la reja del valcón de la Cámara Santa para mostrar la Sancta Sábana y den horden en ello juntamente con el Administrador*".

had its own exhibition particularly during important liturgies. Moreover, it already had significant veneration. In subsequent documents, the Sudarium takes a relevant role and gets the second place in the list of relics just ahead of the large part of the shroud. It is not named as "a part of" but as the whole relic. Since then, the Sudarium has had three specific feasts a year for its veneration: Good Friday, 14 September and 21 September.

The documents that give the Sudarium total prevalence in the second half of the 16th century assume that the Sudarium arrived inside the Holy Ark. Although we are not certain about the route of the Sudarium, the analysis of the pollen grains carried out by Max Frei identified the species of *Acacia albida* which is not recorded in the entire Mediterranean area except in Israel. At least we have this evidence that the Sudarium was in Israel before it arrived in Spain.

If the Sudarium came in this Ark, it was not stored unfolded in its whole dimension inside the big reliquary that was used for its preservation in the Holy Chamber until the 21st century. The folds of the Sudarium show it was folded at least in 8 layers forming a rectangle as reduced as approximately 21 × 27 cm (Fig. 10.1). More creases suggest it was folded in 16 or even 32 layers which turns out to be a size as small as 10 × 14 cm.

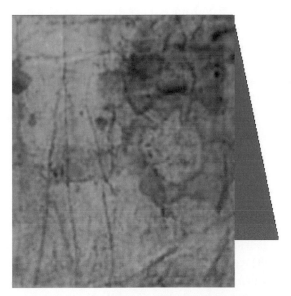

Figure 10.1 Folding of the Sudarium in 8 layers for saving.

Since the end of the 16th century to the beginning of the 21st century the way in which the Sudarium has been preserved and exhibited has hardly changed. It has been protected by velvet, silk and other fabrics and supported by a silvered frame.

In April 2014, the conservation and presentation of the Sudarium changed. It was supported in an inclined position of 30 degrees to reduce the stress of its own weight and without any other canvas covering it. It is in a sealed urn with a Nitrogen atmosphere to prevent the growth of microorganisms. The upper part of this urn or reliquary is made of transparent safety glass. The whole reliquary is enclosed in another larger container that protects it from light with a facsimile that covers the frontal window.

10.1.2   Scientific Study

Since 1965 and thanks to the work of Msgr Ricci, the scientific interest in the Sudarium was aroused and numerous analysis and studies were carried out on it, some directly on it and others in the archives and libraries. Ricci proposed for the first time that the Sudarium had been used to cover the face of Jesus from Golgotha to the tomb and, once there, it was placed separately.

The Research Team of the Spanish Centre of Sindonology (EDICES) continued the investigations of the Sudarium. It meant a decisive contribution to the knowledge of this cloth in Spain and abroad. Among the main efforts of EDICES to teach about the Sudarium is the organization of two Congresses to consolidate the different lines of research. The first Congress was held in 1994 and the second one in 2007.

The Sudarium of Oviedo is made of linen with a size of about 83 × 53 cm. Since it does not present selvage on its edges, it must have been cut from a larger fabric before sale or it was cut at some moment before or after the use on the corpse.

The total weight is about 89 g. It has a texture of taffeta (or tabby). The fabric has a very tight density of 43 × 19 threads/cm (warp × weft) compared to linen textiles manufactured in the Holy Land which usually have 10 × 15 threads/cm. The threads of the Sudarium are twisted in Z while the most frequent spun found in Israel is S-twist. The high density and the spun Z are characteristics in common with the Shroud of Turin indicating a foreign manufacture outside of Israel. A possible country for the manufacture of these burial cloths is India, where the regular spinning was Z[8].

Each thread of the Sudarium is made of between 100 and 120 fibres. The fibres have diameters between 10 and 20 micrometres.

Dr Villalaín and G. Heras provided the sequence of the use of the Sudarium from the cross to the tomb with estimated intervals of time and position of the corpse in each phase. According to the forensic research, the Sudarium was used around the head of a corpse of a man with a beard and moustache, someone who had died in a vertical position. At first, the cloth wrapped the head of the Man after He had stayed in a vertical position for about one hour after his death[9]. The Sudarium was fixed to the lock of the hair behind the nape, and then it was deployed along the left side of the head over the left ear to reach the face and up to the right cheek. The rest of the Sudarium was folded back to cover the face with a second layer. The two layers collected a mixture of blood and oedema in the rate of $1/6$[10] in front of the face. The fluids leaked out from the mouth and nose of the corpse, and they reached the low part of the symmetric bloodiest stains on the two layers. A trustworthy explanation is that the corpse was with the chest in inspiration and with abundant pulmonary oedema, hanging, with the head tilted downwards. As the early and progressive stiffness sets in, the pulmonary oedema begins to flow smoothly through the mouth and nose. A punishment compatible with the scourging would explain the pulmonary oedema that could have increased because of the effort of carrying a weight such as the cross. The left part of the cloth as it is seen in Fig. 10.2 (standard arrangement) shows the side that was in direct contact with the head of the corpse. Figure 10.3 shows a coarse scheme of where the face and the nape were in contact with the cloth.

---

[8]Chapter 3.
[9]Chapter 4.
[10]Chapter 4.

Figure 10.2 Sudarium of Oviedo out of its reliquary (© CES).

Figure 10.3 Scheme of face and nape on the Sudarium.

Since in the Jewish culture, the blood of the deceased must be buried with the body, the Sudarium could be used to contain the blood and to save it to be buried later with the corpse[11].

The Sudarium presented lines of small holes as evidence of seams that fixed the cloth to the hair on the head or beard of the corpse. Some seams correspond to the stitching to the beard.

---

[11]Chapter 3.

Other holes correspond to the nape area. They were used to sew the lock of hair at the back of the head[12].

With the Sudarium fixed firmly, the corpse stayed about an hour in vertical position. The characteristics of the symmetric stains imply that the corpse was later taken down and he was put face down for almost an hour more. In this posture, the upper part of the symmetric stains was soaked. Probably the corpse maintained the original posture of the arms. Then, the Sudarium surrounded the whole head[13].

Subsequently, mobilizations produced several waves of fluid in the central stain originated during the manipulation of the corpse. At some moment, a knot was done at the part which was on the top of the head[14]. If the Sudarium of Oviedo was used for Christ, it remained on his head till the arrival at the tomb. Just before the corpse was covered with the Shroud, the Sudarium was taken off and put aside close to the corpse.

In addition to the bloodiest stains in front of the face, the Sudarium also presents stains corresponding to other wounds, some of them produced by dot shaped objects compatible with a crown of thorns. An hour before the placing of the Sudarium, the dot shaped wounds of the nape were bleeding. Other stains were identified as butterfly, lower left corner, diffuse and stain in accordion[15].

The infrared, ultraviolet, and transparency photographs allowed discriminating the nature and intensity of the blood stains. The stains below the dot shaped wounds ("butterfly and the stain in the corner) placed at the back of the neck have the same properties under infrared light as the stains from the mouth. The origin of these two groups of stains is under discussion. If they come from the mouth and nose, the corpse must have been on its back at some point after the descent from the cross.

On the other hand, under ultraviolet light, the dot shaped wounds show a lighter halo around the darker centre. The "butterfly" and the central stains do not show a halo at the edge of the stain. It supports the fact of considering the different types of blood of

---

[12]Chapter 3.
[13]Chapter 4.
[14]Chapter 3.
[15]Chapter 4.

these stains, vital at the nape and post-mortem around the mouth. The transparency photography helps to determine which one of the post-mortem stains was the first to be in contact with the face and to establish the sequence of their formation[16].

The presence of human blood on the Sudarium of Oviedo has been verified by different Spanish as well as Italian specialists. After a quantitative test, the blood group of the Sudarium must be considered AB according to the possibilities offered by the current forensic haematology. The genetic analysis of a blood sample also provided the presence of human mitochondrial DNA. The sequence of the HVI region only appears in one case over more than 8000 having an important identification value[17].

The presence of blobs of fibrin supports the assumption that the Man of the Sudarium must have undergone severe trauma and previous injuries some hours before the application of the cloth on his head[18].

The Sudarium was sprinkled with powder of aloes and storax. The analyses of specific particles found storax which was a compound that replaced myrrh because it was cheaper, and Palestine was a large producer of it. The aloe applied left its traces mainly in the stains of blood on the surface that was in contact with the face. It implies that the Sudarium was taken off from the head before the applying of the funeral unguents[19]. Pollen grains of *Helichrysum* were found as well, they had got adhered to the cloth when the fluid was still fresh. According to Marzia Boi, the presence of the *Helichrysum* on the Sudarium is the evidence of the funeral use of it. The analysis with fluorescence of X-rays of the dust carried out by the author and his collaborators of EDICES showed a statistically significant higher presence of Calcium particles in the areas with bloody stains. They were probably fixed to the cloth when the physiological fluids were still fresh or soon after. Therefore, this dust comes from the soil where the death of the Man took place. A first appraisal to be verified indicates that this dust corresponds very well with the soil of Calvary[20].

---

[16]Chapter 3.
[17]Chapter 4.
[18]Chapter 4.
[19]Chapter 3.
[20]Chapter 3.

10.1.3 Dating

A piece of the Sudarium was removed in 1986 under request of Msgr Ricci for dating analysis. The two first radiocarbon datings were communicated by Baima Bollone in the 1994 Congress. Several years later, the EDICES cut a new sample and requested in 2007 a new dating that confirmed the previous results: The Carbon 14 dating of the Sudarium of Oviedo suggest the year 748 AD as the most likely for the harvesting of its flax. This dating excludes that the cloth is a relic of Christ. However, other analyses like the Roman crucifixion, the scourging, the knot recorded by Nonnus of Panopolis, the pollen of Israel, the dust of the Calvary, etc. fairly point to an age compatible with the death of Christ.

The age of the linen of the Sudarium was estimated by Raymond Rogers by counting the numbers of crystal defects and the content of vanillin. He concluded that these features of the Sudarium are similar to those of the Shroud in type and quantity and must belong to the Roman period.

Any forgery seems to be quite unlikely[21]. The statistics of the three studies covering more than 300 samples dated by C14 technique shows that about 20% of dating results are rejected by the archaeologists. Whether the radiocarbon dating was wrong or not is one the most prominent open questions about the research on the Sudarium of Oviedo.

10.1.4 Sudarium and Shroud

The Sudarium of Oviedo corresponds very well with the gospel of St John. Its research leads to conclude that the Sudarium was used in the cross and remained until the moment the corpse was put into the sepulchre. The Gospels speak of a Shroud as well. The Shroud of Turin shares with the Sudarium of Oviedo the tradition to be a burial cloth of Christ. If the Shroud of Turin was used for Christ, it covered the corpse instants after the Sudarium of Oviedo was removed from the head of Christ. If this were the case, both cloths must have common features.

We have already mentioned above the common characteristics of their fabrics: spun Z and high density of threads by centimetre square.

---

[21]Chapter 6.

In both cases there is evidence of a crucifixion as the punishment that led to the death of the victim. The crown of thorns left unmistakable traces in both cloths. The crown of thorns was a specific torture of the Passion of Christ. The high geometrical and physiological correspondence of the wounds caused by the thorns cannot be a product of chance. There is good coincidence of shape and distance between wounds in the occipital area. In both cases it is vital blood. Even a clot in the same place of the forehead is shared in both cloths. The central part of the clot is in the Sudarium and the contour of the clot is in the Shroud[22]. While the Roman scourging is evident in the image of the Shroud, this punishment is proposed to be the cause of one mark in the Sudarium.

The geometrical coincidence of the anatomical features such as the beard, moustache, size of nose, etc. is also remarkable. The tuft of hair in the back of the neck left traces in the Shroud as well as in the Sudarium[23]. The latter was stitched to the hair on the back of the head. The hair presumably had dirt on it and was soaked with blood and sweat. After drying for more than an hour, and pressed by the cloth, the hair would have retained the shape of a tress looking like a ponytail which was perceived by some researcher of the Shroud. Therefore, the "ponytail" of the Shroud is explained by the use of the Sudarium on the same corpse[24].

The locks at the sides of the face appear to be running along the cheeks in the image of the Shroud although they should be falling back if the corpse was lying on his back. However, if the Sudarium of Oviedo wrapped the head of the Man of the Shroud, once again the main use of the Sudarium should be considered to be enough to understand why this hair, dirt and soaked with blood and sweat, remained in that position along the cheeks once the Sudarium was removed.

In both cases the blood group is AB. The random probability of this coincidence is about one in a thousand.

The particular and atypical concentration of dust near the tip of the nose on the two cloths is another coincidence which can hardly be explained by chance.

---

[22]Chapter 7.
[23]Chapter 7.
[24]Chapter 8.

The authenticity of both cloths becomes reinforced since their features correspond to the same individual.

10.2 Some Unsolved Questions

10.2.1 Dating

Among the main questions still open in the study of the Sudarium of Oviedo, the issue of the radiocarbon dating stands out above all.

Today, there is no solid scientific evidence that could really challenge the validity of the radiocarbon dating of the Sudarium. There is not a single identified cause that can shift the C14 dating from the first century to the 8th century for the Oviedo's cloth.

Since the samples sent to the laboratories did not have too many foreign elements, contamination by mineral or biological material may not probably be the cause of a wrong dating.

Although the hypothesis that an irradiation by neutrons in the sepulchre could "make the Sudarium and the Shroud younger" is astonishing, it would explain the dating shift in both cases. Passing up the metaphysical viewpoint, we can undertake the physical task and try to exclude or confirm the hypothesis by detection of chlorine 36 isotope. The drawback is that it requires burning tens of milligrams of fabric.

Another approach to answer the irradiation question through non-destructive tests is the analysis of changes in the FTIR (Fourier-transform infrared) spectrum of irradiated linen[25]. The irradiation of an ancient mummy linen implies changes in the region between 1500–1750 cm^{-1}. An increase at the 1640 cm^{-1} peak was observed in the mummy linen together with two new peaks at 1710 cm^{-1} and at 1543 cm^{-1}. A thread off the Sudarium was tested under FTIR. The spectrum of the Sudarium showed the increase at the 1640 cm^{-1} peak wave number with a height that is similar to the peak of the irradiated mummy sample. However, the other two peaks at 1710 cm^{-1} and at 1543 cm^{-1} were not present. The FTIR result for the Sudarium is not enough to draw any conclusion on whether it was exposed to neutron radiation.

---

[25]Chapter 6.

More research of FTIR on ancient and new linen before and after irradiation is needed to verify whether these frequencies are truly linked to neutron irradiation. The research should be addressed to the Shroud as well as the Sudarium. In the case of the Shroud of Turin, the effect must be more remarkable.

Another alternative dating methods based on infrared (FTIR), Raman spectroscopy, and mechanical parameters of single fibres have been tuned by Giulio Fanti. These methods have been applied to the Shroud and provide a dating compatible with the time of Christ. These methods have not been applied to samples of the Sudarium. If some of the already existing cut samples were available, this dating method, once matured, could be applied to the Sudarium because it is almost non-destructive.

Another alternative method that has been evoked for dating is "depolymerisation" or the reduction of the degree of polymerization as a function of aging time. It can be measured by wide-angle X-ray scattering (WAXS). However, this technique is still in the development phase and its main weakness is its dependence on unknown conservation conditions. The main advantage is that this method can be performed non-destructively on a single thread of a few mm long because the tested area is about 0.2×0.5 mm^2. We can say again that this method has not been applied to samples of the Sudarium, but it could be applied, once matured, on existing samples because it is non-destructive.

10.2.2 Islamic Crucifixion

The possibility that the Sudarium of Oviedo were used for a mockery of a Christian by Muslims should be analysed in depth. The Carbon 14 dating can correspond to a crucifixion carried out in Al-Andalus in the South of Spain in the 8th or 9th centuries. Islamic law apparently includes crucifixion and there were crucifixions in Al-Andalus at least in 749 and 813. Therefore, a possibility remains that the Sudarium could belong to a crucifixion by Muslims who had already begun to practice this type of execution in the year 750 in which the relic has been dated. We can assume the crucifixion of a Christian by Muslims in the South of Spain in the second half of the 8th century. We can even assume that the mockery could include the crown of thorns whose traces are also present in the Sudarium. However, details of many of these crucifixions are unknown and when they reach us, they sometimes include remarkable irregularities such as that of a crucified face down in 850[26]. An investigation remains to be carried out about the Islamic and Christian customs in the 8th century. Going in deep in this issue is a task for specialists of the Islamic culture in the past.

To dismiss the possibility of the use of the Sudarium in an Islamic crucifixion, we must consider other features found in the Sudarium which are difficult to conciliate with a Muslim mockery. If we try to recreate the scene, we wonder if the Islamic executioner would allow access to the corpse. More probably, the Muslims would leave the corpse abandoned once they finished the mockery. Then, Christians were allowed to take care of the body. Why would they put a sudarium around the head of the corpse one hour after the death? The easiest hypothesis is that they took the corpse and buried it directly without any previous care. At that time, Christians did not need to bury the blood of the deceased and nor did they use aloe and storax to preserve the blood-stained cloths.

In any case, in the Sudarium there is a possible trace of the Roman scourge tool, the *flagrum*. A couple of small stains identify the punishment undergone by the Man of the Sudarium as a kind of Roman penalty and rules out Islamic crucifixion.

10.2.3   Did the Sudarium Arrive inside the Holy Ark?

The documents listing the relics that came inside the Holy Ark mention "part of the Sudarium". Was this sudarium the one that is venerated today in the middle of the Holy Chamber? The question arises because; in this room are the Sudarium and a part of the shroud. The mention of only one burial cloth, sometimes sudarium and sometime shroud, in the oldest documents, together with the confusion of its meaning between true Sudarium (napkin) and Shroud at those times, prevents us from being sure which of them came in the Ark.

---

[26]Flori, J. (2004). *Guerra Santa, Yihad, Cruzada. Violencia y religión en el cristianismo y el islam*. Biblioteca de Bolsillo, n° 13, Universidad de Granada. pp. 129–130.

The analysis of this subject should answer the following questions:

- How many burial cloths came inside the Holy Ark?
- Could the expression "part of" exclude the whole Sudarium?
- What does mean "sudarium" in the old documents, shroud, or napkin?
- Why was there a silence about the Sudarium for four centuries?

Here we only highlight the use of the word "sudarium" that was, and still is, used to designate the "shroud". Thus, when we read "sudarium" in the list of relics that came inside the Holy Ark, we do not know if this refers to the "napkin" or to the "shroud".

As we saw above, in some texts, a large cloth that would serve to cover the entire body is called a sudarium. But this type of cloth would correspond more specifically to the word "shroud".

According to Guscin[27], the word 'sudarium' was often used to refer to the Shroud, which is a canvas of great proportions. On those occasions, when the expression says, "the Sudarium that covered the head of Jesus *in the sepulchre*", most of the time it is the Shroud. It is the expression "*in the sepulchre*" which makes the difference. In its absence, it would be the small sudarium or napkin.

An ancient text in Latin includes this confusion. In the book *Arculfi relatio de locis sanctis, ab Adamanno scripta*[28], the pilgrim Arculf on his way through Jerusalem around the year 679, describes a dispute between Jews and Christians where a shroud is put to the test by fire:

> *De illo quoque sacrosanto* **sudario** *quod* **in sepulchro** *Domini super caput ipsius fuerat positum. [...] in scrinio ecclesie in alio involutum linteamine recondunt.*

---

[27]Guscin, M. (1998). The Oviedo Cloth. The Lutterworth Press, p. 11 and personal communication. E-mail from Guscin to Barta, 24 October 1998. Subject referencias sudario. Also, Guscin, Mark. 2006. *La Historia del Sudario de Oviedo*. Ayuntamiento de Oviedo. Avilés. p. 57.

[28]Bennett, J. (2001). *Sacred Blood. Sacred Image. The Sudarium of Oviedo*. Libri de Hispania. Colorado. p. 26. Also, Guscin, Mark. 2006. *La Historia del Sudario de Oviedo*. Ayuntamiento de Oviedo. Avilés. pp. 62–64.

Quod noster frater Arculfus alio die de scrinio [e]levatum vidit et inter
populi multitudinem illud osculantis, ipse osculatus est in ecclesiae
conventu, mensuram longitudinis quasi **octonos** *habens* **pedes**[29]

In addition to using the expression "*in the sepulchre*" the
shroud that was removed from the sanctuary measured 8 feet,
that is 2.4 metres.

Another example of the confusion between shroud and
sudarium is the case of the transfer of 22 relics, which were kept
in Constantinople, to Paris. Between 1239 and 1242 the Latin
emperor Baldwin II sent these relics to his relative Luis IX
(Saint Louis) who housed them in *Saint Chapelle*[30]. In the list of
relics, the item at position 16 is:

Partem sudarii qua involutum fuit corpus ejus in sepulchre:

The mention of the sepulchre and the whole body implies that
it is part of the shroud.

Maybe the most significant example is the famous controversy
of Bishop Pierre d'Arcis. Around 1389 in the second exhibition
of the Shroud of Turin, d'Arcis named it in Latin "sudarium" which
was clearly the Shroud today in Turin.

in ostensione dicti **sudarii**, *quod domini* **sudarium** *ab omnibus*
credebatur

Likewise, the Anti-Pope Clement VII in his bulls to regulate the
matter also calls it the sudarium.

non est verum **Sudarium** *Domini nostri Jhesu Xpisti, sed quedam*
pictura seu tabula facta in figuram seu representacionem **Sudarii**,
quod fore dicitur ejusdem Domini nostri Jhesu Xpisti.

Even today, people in Spanish America often use "sudario"
for the Shroud of Turin.

It is significant for our discussion that in the Chapter Minutes
of 17 September 1557[31], the Sudarium is named holy shroud

---

[29]Arculfi relatio de locis sanctis scripta ab Adamnano. c670. I. Pomialovky. Book 1,
XI, pp. 11–14.

[30]Durand, J. (2001). Le Tresor de la Sainte Chapele. Paris. Louvre.

[31]López Fernández, E. (2016). Historical sources about the relics of Oviedo's
cathedral. Territorio, Sociedad y Poder, 11. p. 15. The minutes are in the Archives
of the Cathedral, *Acuerdos capitulares*, Libro 8, f. 586v. "*Cometieron a los ss. [] para*
que vean cómo se ha de hazer la reja del valcón de la Cámara Santa para mostrar
la Sancta Sábana y den horden en ello juntamente con el Administrador".

("*sancta sabana*")[32]. It is a clear example from that time of the fact of mixing up of the names of the two relics that today we want to differentiate.

According to Enrique López, once the Sudarium was "discovered" in the middle of the 16th century for several years and at least until 1574, in the references, the holy Sudarium is often named holy shroud. This mistake seems to be shared in 1572 by Ambrosio de Morales when he visited the Cathedral[33].

As we said, the whole Sudarium and a part of the shroud that measures approximately 25 × 25 cm are preserved today in Oviedo. In the Langreo document, which is the oldest with the list of relics, the "part of the *sudario*" appeared in the 9th order and no other burial cloth was included in the list. Which burial cloth was it referring to? In the Valenciennes 99 document, it is stated that the relics of the Holy Ark were found inside their boxes each one with its label in Latin[34]. We can assume that the label of a relic said "*sudario*" and it was included as that in the list. According to the Langreo list, no label said "*sindone* (shroud)". However, several subsequent documents after the first, said "part of the *sindone* (shroud)" in the place of the Sudarium. Sometimes only *sudarium* is included, sometimes only *sindone* is included, and sometimes both are included. Why this change of name? The known documents allow us to assume that there was no new opening of the Ark before the change of name. The change could not be because the reading of a forgotten label. Pelayo was the first using "sindone" instead of "sudario". Did he think both terms referred to the same cloth?

As an example of the change, in the bulletins (sheet for pilgrims) of the 15th century, among the first group in the list of relics is a part of shroud (*de sindone*) but there is no mention of the Sudarium in the whole list[35].

It is probable that the "part of the sudarium" included among the relics of the Ark was the part of the shroud present until today in the cathedral and the whole Sudarium was another cloth.

[32]Chapter 2.

[33]López Fernández, E. (2007). *La Veneración del Sudario de Oviedo a través de la Historia*. Oviedo, Relicario de la Cristiandad. Actas del II Congreso Internacional sobre el Sudario de Oviedo. p. 361.

[34]López Fernández, E. (2004). *El Santo Sudario de Oviedo*. MADU. Granda-Siero. p. 44. footnote 50.

[35]López Fernández, E. (2018). *El Santo Sudario de Oviedo*. Soluciones Gráficas. p. 251.

If we doubt about which of the two cloths the sources refer when they only mention one of the burial cloths, we can opt for the part of the shroud because the particle "*de*" corresponds better to the small piece of shroud than the whole Sudarium. We cannot exclude the hypothesis that the Sudarium was never inside the Holy Ark. If the mention refers to the part of the shroud, when did the Sudarium arrive in Oviedo? Only since the second half of the 16th century the Sudarium is clearly referred to, together with its particular exhibitions and great prominence.

Since several sources such as old inventories did not mention the Sudarium nor the shroud, it follows that they were not considered objects of capital importance[36]. We could safely affirm that if the Sudarium was inside the Holy Ark, it was not deployed nor was it observed in its full size for four centuries.

With this state of knowledge, we should research more about the possibility of another way for the Sudarium to arrive in Oviedo, maybe some time before the second half of the 16th century. The reliable documented history of the Sudarium of Oviedo starts about that time.

An experimental test that can help to elucidate this issue is the dating of the wood of the frame and the box that were used to preserve the Sudarium for centuries. These have now been withdrawn to use the modern preservation methods as recommended by EDICES. It allows knowing when the Sudarium was placed in these reliquaries.

Other odd issue is the homogenous silver contamination over the entire surface of the Sudarium[37] would point to a silver reliquary large enough to keep the relic totally deployed. Such a rich hypothetical reliquary could not have weirdly disappeared. However, there is no documented reference to such a box in Oviedo.

If the Sudarium did not come to Oviedo before 1075, a reference can get relevance. At the end of the 12th century sudarium and shroud are mentioned simultaneously in Constantinople in the Pharos chapel[38]. Nicholas Soemundarson saw both in his 1157 visit. This sudarium was not sent to Paris. If it were the

[36]Nicolotti, A. (2016). *The Shroud of Oviedo: Ancient and Modern History*. Territorio, Sociedad y Poder. Revista de estudios Medievales. N. 11.
[37]Chapter 3.
[38]Chapter 2.

same sudarium preserved today in Oviedo, it could have left Constantinople before or after the 4th Crusade. Pilgrims visiting Constantinople after the Fourth Crusade do not mention a sudarium among the relics remaining in the city. This makes possible that the sudarium having left the city. However, there is no evidence that the sudarium did still not remain there.

10.2.4   The Occipital Stains of the Sudarium

The origin of the corner stain that is close to the butterfly stain is an issue for discussion. Dr Sánchez Hermosilla and other members of the EDICES claim that this stain could have its origin in a wound on the back of the Man where the spear exited when the crucified was run through. In this case, and assuming the common use of the Sudarium and Shroud for the same corpse, we would expect a confirmation in the image of the Shroud of Turin. However, there is no evidence of such an exit wound in the Italian cloth. Moreover, we can estimate where the corner stain falls when we superimpose the dot shaped stains corresponding to the crown of thorns over both cloths. When the maximal fitting is assumed, the corner of the Sudarium falls to the back of the neck near the cervical vertebras. Fig. 10.4 shows the result of this case identified as Position 1. This position is too high to correspond to a wound of the spear coming out through the back of the Man of the Shroud. However, in this case, the head corresponding to the distance between the occipital wounds and the tip of the nose result a too narrow head[39]. This difficulty disappears when we realize that the dot shaped stains were in the hair and not in the scalp. The stitched hair tress shifted from its central position to the left, approaching the left ear when the Sudarium was tightened around the face.

Dr Juan M. Miñarro carried out the reconstruction of the 3D body of the Man of the Shroud. For the head, he used the data of the Sudarium as well. In the Spanish journal Linteum[40], the accurate study was developed. He used the anatomical points

---

[39]Chapter 9.

[40]Miñarro, J. M. (2014). *Sobre la Compatibilidad de la Sindone y el Sudario*. Linteum 56. January-June 2014. pp. 4–56 (Monograph).

of the skull to place the wounds of the occipital area of the Sudarium over the corresponding area of the Shroud. This gave a different superposition with respect to the above mentioned. The result is shown in Fig. 10.4 is identified as Position 2.

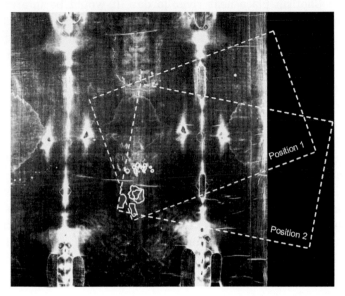

Figure 10.4 Place of the stain of the corner depending on the fitting of different features.

In this case, the corner of the Sudarium falls higher than in the Position 1 and shifted to the right side of the neck. It seems to be very unlikely that this stain corresponds to a wound of the spear. The characteristic of the stains seems to have the same origin as the stains from the nose and mouth.

Another open question related to the area of the Sudarium corresponding to the nape is the following. Although the way of using of the Sudarium would imply a tuft of hair in the back of the neck, the direction of the fold that runs across the middle of the "butterfly" stain rather indicates that the cloth did not wrap the tuff of hair as expected, passing over the tress and hiding the edge of the Sudarium between the hair and the neck. Moreover, the presence of the ponytail in the Shroud of Turin is dismissed by some researchers. They think it is an effect of the image.

10.2.5   Other Miscellaneous Open Issues

As far as it concerns the old history of the Sudarium, it remains a challenge for future research to update the findings in the caves at the Baptism place near river Jordan. It would be interesting to know if the tombs found by the archaeologists have skeletons of the seven virgins mentioned by the pilgrims of Piacenza. If the place is confirmed, samples of the site can allow further analysis of compatibility with the dust of the Sudarium of Oviedo.

As for the places where the Sudarium stayed or was kept, we can use the pollen as geographical indicators. The results provided by Max Frei should be completed with identification of the species of pollen found in the Sudarium and compare it with the endemic species of *Pistacia palestina* and some species of *Tamarix*, which are endemic or native Palestinian species.

The test of the X-ray fluorescence did not take into account the impact on the result of the table used as a support. To assess the possible origin of the dust, which is present on the Sudarium, additional tests should be done. Measures of the samples of the Calvary and of the Cathedral should be performed in the same conditions as the measured in the Sudarium to allow reliable results.

In addition, the iron distribution could be correlated with the intensity of the blood stains in the Sudarium if the measurement is performed in absence of any support behind the cloth.

The high density of weave of Sudarium and Shroud could be compared to the fabrics of other times and geographical places to reduce the possible origin of their manufacturing. As a first approach, we find that both cloths have more threads in one square centimetre than the fabrics of the 16th and 17th centuries around the Alps[41].

One open question that rose in this book was for another relic. It deals with the Tunic of Argenteuil. If the spun is Z, it fits the spun of the part of the Tunic preserved in the Cathedral of Toledo. This last came from the *Saint Chapelle.* And the part of the Tunic of the church of Paris was previously in Constantinople. It could be worth verifying the possible compatibility of both cloths. In the case that the Tunic of Argenteuil and part of the Tunic coming from Constantinople and saved in Toledo are the same,

---

[41]Chapter 3.

the simultaneous presence of the Tunic in Argenteuil and one of its pieces in Constantinople before 1250 would need to be researched.

10.3 The Sudarium as a Means of Evangelization

The interest for the Sudarium is not only for scientific investigation searching an answer to the question of its authenticity. It can be a way to evangelize. The chapter of canons of the Cathedral of Oviedo is focused on this aspect of the assumed relic of Christ. The studies that support the authenticity promote the veneration. It allows priests and bishops to bring the Lord's Passion closer to the faithful and recreate it.

A good example is the testimony of Janice Bennett when she contemplated the Sudarium in the sacristy of the Cathedral of Oviedo. The Dean-President of the Chapter of the Cathedral, D. Rafael Somoano Berdasco told her [The Sudarium] "is the heart of Jesus and the sign of His love for us."[42]

Now, independently from the authenticity question, the studies of the scourging and crucifixion process allow us to know the harsh reality of the sufferings of Christ. The possible authenticity of the Sudarium provides us with the details of events between the death on the Cross and the placement in the sepulchre that the Gospel does not give us. Believers want to know every detail about their lord.

"The contemplation of the bloodstains, for people who are predominantly concerned with material comforts, helps us to penetrate the meaning of suffering and death. The thought that the Son of God willingly subjected himself to the total powerlessness of death lead us to discover the mystery of suffering that achieves salvation for all humanity, in the words of Pope John Paul II."[43]

Some people draw near the Sudarium, in the main altar on its feasts and in the Holy Chamber during the rest of the year,

---

[42]Bennett, J. (2001). *Sacred Blood. Sacred Image. The Sudarium of Oviedo.* Libri de Hispania. Colorado. p. 187.

[43]Bennett, J. (2001). *Sacred Blood. Sacred Image. The Sudarium of Oviedo.* Libri de Hispania. Colorado. p. 187.

in order to pray and to be compassionate with their God. This type of pilgrims would like to see the Sudarium permanently. They are disappointed when the time the Sudarium remains exhibited in front of the altar is not enough to be venerated by the whole crowd and it is removed while the line of people approaching the cloth is still long. Curators have to balance the desires of these worshipers and prioritize its preservation for future generations. The Chapter of canons gives reasonable priority to the safety of the Sudarium over the desire to permit the public to see it often. They keep the tradition followed for centuries to show it in the three established events and on few exceptional occasions.

The lovers of the Lord and pilgrims coming to the Cathedral have to settle for a facsimile in spite of their desires. They would like to see the true Sudarium at a few centimetres, to touch it and to kiss it. In the current preservation conditions, it is exceptional to see it and almost impossible to touch or kiss it. This shows that the Sudarium has never been exploited to attract pilgrims.

The indulgences associated with the Sudarium of Oviedo mean that it can be a way to access Heaven. There are innumerable reasons why the Sudarium of Oviedo is significant for the world today. It is the historical proof that Jesus indeed lived on this earth and died.

It confirms and updates the reality of the Passion and opens the mystery to us. There is no doubt that in the future, it will bring believers and scientists closer to the truth.

Name Index

Subject Index